Lecture Notes in Computer Science 12115

Yunmook Nah · Chulyun Kim ·
Seon Ho Kim · Yang-Sae Moon ·
Steven Euijong Whang (Eds.)

Database Systems for Advanced Applications

DASFAA 2020 International Workshops

BDMS, SeCoP, BDQM, GDMA, and AIDE
Jeju, South Korea, September 24–27, 2020
Proceedings

 Springer

Editors
Yunmook Nah
Dankook University
Yongin, Korea (Republic of)

Seon Ho Kim
Integrated Media Systems Center
University of Southern California
Los Angeles, USA

Steven Euijong Whang ⓘ
Korea Advanced Institute of Science
and Technology
Daejeon, Korea (Republic of)

Chulyun Kim
Department of IT Engineering
Sookmyung Women's University
Seoul, Korea (Republic of)

Yang-Sae Moon ⓘ
Kangwon National University
Chunchon, Korea (Republic of)

ISSN 0302-9743 ISSN 1611-3349 (electronic)
Lecture Notes in Computer Science
ISBN 978-3-030-59412-1 ISBN 978-3-030-59413-8 (eBook)
https://doi.org/10.1007/978-3-030-59413-8

LNCS Sublibrary: SL3 – Information Systems and Applications, incl. Internet/Web, and HCI

This Springer imprint is published by the registered company Springer Nature Switzerland AG
The registered company address is: Gewerbestrasse 11, 6330 Cham, Switzerland

Preface

Along with the main conference, the DASFAA 2020 workshops provided international forums for researchers and practitioners to introduce and discuss research results and open problems, aiming at more focused problem domains in database. This year, five workshops were held in conjunction with the 25th International Conference on Database Systems for Advanced Applications (DASFAA 2020):

- The 7th International Workshop on Big Data Management and Service (BDMS 2020)
- The 6th International Symposium on Semantic Computing and Personalization (SeCoP 2020)
- The 5th Big Data Quality Management (BDQM 2020)
- The 4th International Workshop on Graph Data Management and Analysis (GDMA 2020)
- The First International Workshop on Artificial Intelligence for Data Engineering (AIDE 2020)

All the workshops were selected through a public call-for-proposals process, and each of them focused on a specific area that contributes to the main themes of DASFAA 2020. Once the proposals were accepted, each workshop was proceeded with its own call for papers and reviewing of the submissions. In total, 22 papers were accepted, including 5 papers for BDMS 2020, 5 papers for SeCoP 2020, 4 papers for BDQM 2020, 4 papers for GDMA 2020, and 4 papers for AIDE 2020. All the insightful papers were presented at the workshops in conjunction with DASFAA 2020, held during September 24–27, 2020, in Jeju, South Korea. The workshops were originally scheduled for May 21–24, 2020, but unavoidably postponed due to the COVID-19 pandemic all over the world.

We would like to thank all of the members of the Workshop Organizing Committees, along with their Program Committee members, for their tremendous effort in making the DASFAA 2020 workshops a success. In addition, we are grateful to the main conference organizers for their generous support as well as their efforts in including the workshop papers in the proceedings series. Lastly, we acknowledge the generous financial support from IITP[1], Dankook University SW Centric University Project Office, DKU RICT, OKESTRO, SUNJESOFT, KISTI, LG CNS, INZENT, Begas, SK Broadband, MTDATA, WAVUS, SELIMTSG and Springer.

September 2020

Chulyun Kim
Seon Ho Kim

---

[1] Institute of Information & communications Technology Planning & Evaluation (IITP) grant funded by the Korea government (MSIT) (No. 2020-0-01356, 25th International Conference on Database Systems for Advanced Applications (DASFAA)).

Organization

BDMS 2020 Organization

Organizers

Xiaoling Wang	East China Normal University, China
Kai Zheng	University of Electronic Science and Technology of China, China
An Liu	Soochow University, China

SeCoP 2020 Organization

Honorary Co-chairs

Qing Li	The Hong Kong Polytechnic University, Hong Kong
Fu Lee Wang	The Open University of Hong Kong, Hong Kong

General Co-chairs

Gary Cheng	The Education University of Hong Kong, Hong Kong
Tianyong Hao	South China Normal University, China
Tak-Lam Wong	Douglas College, Canada

Organizing Co-chairs

Di Zou	The Education University of Hong Kong, Hong Kong
Winnie Wai Man Lam	The Education University of Hong Kong, Hong Kong

Technical Program Committee Co-chairs

Yi Cai	South China University of Technology, China
Wei Chen	Agricultural Information Institute of CAS, China
Leonard Kin Man Poon	The Education University of Hong Kong, Hong Kong
Haoran Xie	Lingnan University, Hong Kong

Publicity Co-chairs

Mingqiang Wei	The Open University of Hong Kong, Hong Kong
Guanliang Chen	Monash University, Australia
Shan Wang	Macau University, Macau
Yu Song	South China Normal University, China
Yanjie Song	The Education University of Hong Kong, Hong Kong

Media Chair

Man Ho Alpha Ling	The Education University of Hong Kong, Hong Kong

PC Members

Zhiwen Yu	South China University of Technology, China
Jian Chen	South China University of Technology, China
Raymong Y. K. Lau	City University of Hong Kong, Hong Kong
Rong Pan	Sun Yat-sen University, China
Yunjun Gao	Zhejiang University, China
Shaojie Qiao	Southwest Jiaotong University, China
Jianke Zhu	Zhejiang University, China
Neil Y. Yen	University of Aizu, Japan
Derong Shen	Northeastern University, China
Jing Yang	Research Center on Fictitious Economy & Data Science CAS, China
Wen Wu	Hong Kong Baptist University, Hong Kong
Raymong Wong	Hong Kong University of Science and Technology, Hong Kong
Cui Wenjuan	CAS, China
Xiaodong Li	Hohai University, China
Xiangping Zhai	Nanjing University of Aeronautics and Astronautics, China
Xu Wang	Shenzhen University, China
Ran Wang	Shenzhen University, China
Debby Dan Wang	National University of Singapore, Singapore
Jianming Lv	South China University of Technology, China
Tao Wang	University of Southampton, UK
Jingjing Wang	The Open University of Hong Kong, Hong Kong
Dingkun Zhu	The Open University of Hong Kong, Hong Kong
Zongxi Li	City University of Hong Kong, Hong Kong
Xudong Kang	The Education University of Hong Kong, Hong Kong
Xieling Chen	The Education University of Hong Kong, Hong Kong

BDQM 2020 Organization

Workshop Co-chairs

Yajun Yang	Tianjin University, China
Wenjie Zhang	The University of New South Wales, Australia

PC Members

Zhifeng Bao	RMIT University, Australia
Laure Berti-Equille	Hamad Bin Khalifa University, Qatar
Yingyi Bu	Couchbase, USA
Gao Cong	Nanyang Technological University, Singapore
Yunpeng Chai	Renmin University of China, China
Qun Chen	Northwestern Polytechnical University, China
Yueguo Chen	Renmin University of China, China
Yongfeng Dong	Hebei University of Technology, China

Rihan Hai	Lehrstuhl Informatik 5, Germany
Cheqing Jin	East China Normal University, China
Guoliang Li	Tsinghua University, China
Lingli Li	Heilongjiang University, China
Hailong Liu	Northwestern Polytechnical University, China
Xianmin Liu	Harbin Institute of Technology, China
Xueli Liu	Tianjin University, China
Zhijing Qin	Pinterest, USA
Xuguang Ren	Inception Institute of Artificial Intelligence, UAE
Chuitian Rong	Tianjin Polytechnic University, China
Hongzhi Wang	Harbin Institute of Technology, China
Jiannan Wang	Simon Fraser University, Canada
Xin Wang	Tianjin University, China
Jianye Yang	Hunan University, China
Xiaochun Yang	Northeast University, China
Dan Yin	Harbin Engineering University, China

GDMA 2020 Organization

Workshop Chair

Lei Zou	Peking University, China

PC-Chairs

Xiaowang Zhang	Tianjin University, China
Liang Hong	Wuhan University, China

AIDE 2020 Organization

Organizers

Chulyun Kim	Sookmyung Women's University, South Korea

Program Co-chairs

Heejun Chae	Sookmyung Women's University, South Korea
Younghoon Kim	Hanyang University, South Korea
Sangjun Lee	Soongsil University, South Korea

Financial Sponsors

Academic Sponsors

Contents

The 5th Big Data Quality Management (BDQM 2020)

The 4th International Workshop on Graph Data Management and Analysis (GDMA 2020)

The 1st International Workshop on Artificial Intelligence for Data Engineering (AIDE 2020)

Low Rank Communication for Federated Learning

Huachi Zhou$^{(\boxtimes)}$, Junhong Cheng, Xiangfeng Wang, and Bo Jin

East China Normal University, Shanghai 200062, People's Republic of China
{hczhou,jhcheng}@stu.ecnu.edu.cn,
{xfwang,bjin}@cs.ecnu.edu.cn

Abstract. Federated learning (FL) aims to learn a model with privacy protection through a distributed scheme over many clients. In FL, an important problem is to reduce the transmission quantity between clients and parameter server during gradient uploading. Because FL environment is not stable and requires enough client responses to be collected within a certain period of time, traditional model compression practices are not entirely suitable for FL setting. For instance, both design of the low-rank filter and the algorithm used to pursue sparse neural network generally need to perform more training rounds locally to ensure that the accuracy of model is not excessively lost. To breakthrough transmission bottleneck, we propose low rank communication Fedlr to compress whole neural network in clients reporting phase. Our innovation is to propose the concept of optimal compression rate. In addition, two measures are introduced to make up accuracy loss caused by truncation: training low rank parameter matrix and using iterative averaging. The algorithm is verified by experimental evaluation on public datasets. In particular, CNN model parameters training on the MNIST dataset can be compressed 32 times and lose only 2% of accuracy.

Keywords: Federated learning · Convolutional neural network · Low rank approximation · Matrix compression · Singluar vaue decomposition

1 Introduction

The widespread application of deep neural networks has achieved remarkable success in many computer tasks, such as image processing, speech recognition, and text translation. However, most of them require people to share their personal data with service providers, including their daily behaviors, personal preferences, etc., which to some extent is quite private data. Therefore, users hate or even refuse to upload personal data to the server.

At the same time, mobile phones and tablets with a large amount of users' sensitive data become more and more powerful. With development of edge computing, the scheme of deploying tasks at the edge of the network rather than the cloud is becoming more and more mature.

© Springer Nature Switzerland AG 2020
Y. Nah et al. (Eds.): DASFAA 2020 Workshops, LNCS 12115, pp. 1–16, 2020.
https://doi.org/10.1007/978-3-030-59413-8_1

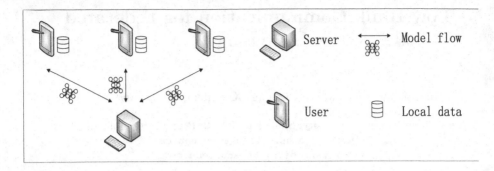

Fig. 1. The structure of FL

FL is a machine learning framework that aims to build a high-quality learning model as centralized in which training data remain local. In order to do this, H. Brendan et al [1] propose Federated Average algorithm. Terminal devices train a shared model with data stored locally under the coordination of parameter server. At each training round, sever randomly selects those volunteering devices which are qualified to participate in this round. Those chosen clients upload their own gradients to the central sever within a fixed time. Then server averages all participants' gradients and sends it back to the participants. This iterative training process continues throughout the network until global convergence is reached or some termination condition is met [1]. However, in the learning process of mass clients, communication efficiency is of the utmost importance (Fig. 1).

On one hand, because of characteristics of asymmetric digital subscriber line, the upload speed is several times lower than the download speed. Network communication speeds are even orders of magnitude slower than distributed environment. In order to solve transmission bottleneck, reducing communication costs, especially when clients upload their models, helps nodes spend less time on crowded channels to speed up training round.

On the other hand, existing work trying to cut communication cost mostly focuses on two schemes. The first is to reduce the amount of information needed to be exchanged globally by increasing the number of local iteration rounds [2]. However, the size of model parameters needed to be transmitted for a single time is still very large. The second is to quantify information to reduce the number of bits [3,4]. But, the problem of quantization is that the accuracy is unstable and it is easy to cause precision loss.

To solve these problems, we propose Federated low rank(Fedlr) algorithm which reduces the communication cost between server and clients, while keeping the accuracy. Our work shows that the low-rank representation of a matrix can optimally achieve a balance in accuracy loss, which brings new ideas to the design of communication efficiency strategy. Our contributions are summarized as follows:

- We apply truncated SVD to compress parameter matrix of model convolutional layers and propose a general algorithm for determining the optimal compression ratio of the parameter matrix.
- In order not to significantly increase model loss caused by rank truncation, we train low rank convolutional layers. The central node aggregates historical gradients of each node separately to reduce the variance of current gradient.
- The extensive experiments on public datasets show the effectiveness of the proposed algorithm.

2   Related Works

In Sects. 2.1 and 2.2, we introduce traditional neural network compression techniques. In Sect. 2.3, we introduce gradient compression techniques. In our description, readers will find that content in Sect. 2.3 is more suitable for FL.

2.1   Model Pruning

The purpose of neural network pruning is to reduce the amount of model weights to make model thinner. There are three distinguished mechanisms. The first is to add different penalty terms to the loss function of neural network during training process to promote the trained neural network to contain more zero elements. [5] selects parameter index set whose value range is {0,1} to penalize the total number of network parameters. [6] sets a scaling factor which multiplies with a channel output before it goes into next layer. The penalty term chooses the sum of scaling factors. The second method directly deletes smaller weights during training. [7] proposes a gentle gradual pruning method, which slowly prunes from a small sparsity to the desired sparsity. [8] inherits this idea and proposes one-shot pruning, while each round masking some smaller weights, resetting remaining weights to initial value and retraining the whole network. In the forward propagation process, [9] deletes a channel of the i-th layer and see if the output of the i+1 layer is seriously affected to determine the importance of the channel weight in i-th layer. The third method is weight sharing: through weight clustering, channels of the same category share the same weight [10], which only decreases the memory space occupied by the model and does not save model reasoning time, compared to the first two methods. Overall, pruning is not suitable for FL environments. First, it is difficult to agree on whether to cut off some connectivity between neurons among clients. Second, pruning needs extra rounds for retraining to restore the performance of the model, which takes a lot of time.

2.2   Model Low Rank Filter and Model Quantization

Another method of model compression is to use low rank convolutional filters. [11] designs low rank filters including two processes. First, it classifies input

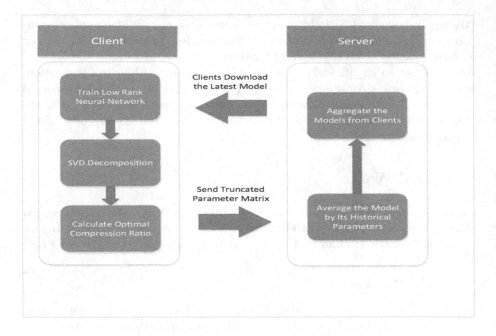

Fig. 2. Flow chart of Fedlr algorithm

channels and only allows portion of filters to connect to a specific class to reduce filters redundancy and projects the input image dimension down to 1D subspace using intermediate filters to reduce input size. Then they apply the tensor decomposition method to all filters to store weights more efficiently. [12] In order to reduce channel or filter redundancy, horizontal and vertical filters with a rank of 1 are used to approximate the previous 2d tensor. The aim of model quantization is to replace floating-point numbers with 8-bit or 16-bit integers or other low-precision numeric formats. Based on [13], there are two quantization methods: deterministic quantization and stochastic quantization. For the former, the popular binary neural network maps model weights to $\{+1, -1\}$ according to their positive and negative. For the latter, [14] assumes model weights follow multinomial distribution and attempts to train neural networks with discrete weights. But under FL setting, low rank filter design also needs additional local training rounds to minimize reconstruction errors, which takes a lot of time. For quantization, decreasing number of bits may cause serious degradation of the model. According to Deep Compression's survey, quantized convolutional layer requires 10 bits to avoid a significant loss of accuracy [15].

2.3   Gradient Quantization

Gradient quantization, just as its name suggests, is applied to the output gradient to be transmitted by each node, which follows a common principle. The mathematical expectation of quantified variable is an unbiased estimate of the original variable mathematical expectation [16–18]. Unlike model quantization, which recovers the accuracy of pruned model by increasing training round, gradient quantization uses other strategies to bridge the gap between quantized values and true values of gradients. [19] Each node accumulates its own quantization error for each round. Before each round starts quantization, decayed quantization error is added to current local gradient to compensate it. [20] uses stochastic rotated quantization to limit quantization range to $[0, 1]$ to reduce mean squared error. Due to its wide application in distributed learning, gradient quantization is also widely used in FL. It is worth noting that our work does not include any distributed quantization and the subsampling techniques contained in [21]. Future work will try to make them compatible with our work to provide greater compression rates (Fig. 3).

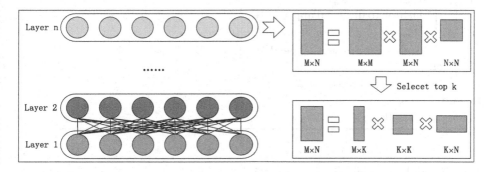

Fig. 3. The left part is n fully connected layers and we assume top layer has shape $M \times N$. Right part illustrates the compression process of SVD

3   Federated Low Rank Algorithm

3.1   Overview

In this section, we will describe our method. The algorithm has four main components: training low rank parameter matrix, determining optimal compression ratio, compressing model using SVD and reducing noise. The complete procedures at the server and each client node are presented in Algorithms 1 and 2, respectively. In order to explain our Fedlr strategy more clearly, we provide a flow chart (shown in Fig. 2) of the algorithm. In left part, clients download latest model, train low rank model locally and upload their model using truncated SVD. In right part, server receives each client model, averages with its historical model and aggregates all averaged models.

Algorithm 1: Fedlr strategy for server:

Input : specified minimum receiving client gradient number T, Specified communication round N, Model selected for training $\widetilde{W}$, volunteering edge nodes set C

Output: shared model parameter W_{out}

1 Initialize $\widetilde{W}$ randomly ;
2 Initialize $history\_gradients \leftarrow []$;
3 Initialize $aggregated\_gradients \leftarrow []$;
4 **for** *each communication round r = 1,2... N* **do**
5   Randomly select S_t clients from C;
6   send $\widetilde{W}$ to S_t clients;
7   **for** $t = 1, 2...T$ **do**
8   listening clients m in S_t and receiving message $\widetilde{U_m}\widetilde{\Sigma_m}\widetilde{V_m}$;
9          $W_m = \widetilde{U_m}\widetilde{\Sigma_m}\widetilde{V_m}$;
10   history_gradients[m].append (W_m) ;
11         $\widetilde{W_m} = \sum_{i=1}^{3} history\_gradients[m][i]$;
12   aggregated_gradients.append($\widetilde{W_m}$);
13   **end**
14     $\widetilde{W} = \Sigma_{i=1}^{T} aggregated\_gradients[i]$;
15 **end**
16 return $\widetilde{W}$ as W_{out};

3.2   Model Compression Using SVD

Low-rank decomposition has a wide range of applications in image restoration, matrix filling, and collaborative filtering. Projecting matrices to lower-dimensional linear subspaces greatly reduces costs of federal learning communication [22]. The adversarial training of Peter Langenberg and others on the fully-convolutional neural network shows that low-rankness of neural network weights can further improve robustness [23]. Low rank characteristic also helps model from overfitting. Compared to other popular matrix factorization techniques, such as CUR matrix factorization, which maintains sparsity within the decomposed matrix, we adopt a more popular truncated SVD matrix technique. There are two advantages. First, according to Eckart-Young theorem, truncated SVD has the smallest low-rank approximation error, which lets us not worry about excessive decline in accuracy. Second, according to extensive experiments, only around 1% of singular values contain 97% of energy, which also explains from another perspective why the SVD low rank approximation still maintains excellent accuracy even if a large number of singular vectors are discarded. This feature of focusing energy on a few singular values allows us to compress the parameter matrix as much as possible. In this paper, we only focus on parameter matrix of convolutional layer and do not consider structure or weight of a single filter in particular. In results analysis section, we will discuss connection between our strategy and single convolution filter. The weight of CNN layer

---

Algorithm 2: Fedlr strategy for clients:

Input : shared model parameter from server W, local datasets D,
minibatchsize b, learning rate r, epoch size T, hyperparameter λ α,
loss function l, Frobenius norm constraint F, clients C
Output: model parameter $U\Sigma V$

1 training phase;
2 **for** *each client* $c \in C$ *in parallel* **do**
3 | Initialize $W_{n,c} \leftarrow W$;
4 | **for** $m = 1,2...T$ **do**
5 | | **for** $n = 1,2...\frac{D}{b}$ **do**
6 | | | $W_{n,c} = W_{n-1,c} - r(\nabla l + \alpha \nabla F)$;
7 | | **end**
8 | **end**
9 | $\sum_{i=1}^{r} U\Sigma V = SVD(W_c)$;
10 | Estimate $k \leftarrow optimal\_ratio(\Sigma)$ from equation (14);
11 | Sending $\widetilde{U}\widetilde{\Sigma}\widetilde{V}$ to server;
12 **end**

---

is in the form of 4d tensor $W^{d \times d \times m \times n}$ (d is kernel size, m is the number of input channel and n is the number of output feature map). We convert W into $d^2 \times m \times n$ shape. So each element in 3d array is $W \in R^{m \times n}$. According to the definition of SVD [24], $W = U\Sigma V$, where U is a left singular matrix with the size of $m \times \min(m, n)$, V is a right singular matrix with the size $\min(m, n) \times n$ and Σ is a diagonal matrix of singular values like $diag(\sigma_1, \sigma_2......, \sigma_n)$ where $\sigma_1 \geq \sigma_2...... \geq \sigma_{\min(m,n)}$. According to the definition of SVD, W can be written as

$$W = \sum_{i=1}^{\min(m,n)} \sigma_i U_i V_i . \tag{1}$$

Assuming that only k ranks are retained, the optimal rank k approximation of the parameter matrix W is:

$$\widetilde{W} = \sum_{i=1}^{k} \sigma_i U_i V_i . \tag{2}$$

The size of original parameter matrix $B = m \times n$. After the parameter matrix is compressed, message sent by a client to the server equals to $\widetilde{B}$ (We don't count the packet header and tail added for network transmission):

$$\widetilde{B} = \widetilde{U}\widetilde{\Sigma}\widetilde{V} = m \times k + k + n \times k = k(m + n + 1) . \tag{3}$$

In order to achieve message compression, the reserved rank number k needs to satisfy $0 \leq \widetilde{B} \leq B$, which means

$$0 \leq k \leq \frac{mn}{m+n+1} . \tag{4}$$

We operate each element in 3d array of all convolutional layers in same way. When the channel is congested and packet loss occurs frequently, mobile phone detects that network layer is in a bad condition with high delay. Instead of sending entire low rank matrix at once, the client sends one singular value and the corresponding left and right singular vector each time.

3.3   Optimal Compression Ratio

In order to reduce redundancy between filters, it is really effective to replace them with a linear combination of fewer filters or low-rank filters. In practical applications, it is very difficult to decide how many filters to be left in order to achieve the maximum compression ratio while a small performance degradation. Researchers face the same problem when compressing neural parameters to be transmitted. In the past, model compression often selects fixed-rank intercep-tion [25]. We suppppose that different machine learning tasks and corresponding datasets demand different parameter space. The result is that the number of ranks worth retaining for different task or different layer is also different. The results analysis section verifies our assumption. When fixed compression ratio is large, information originally contained in the parameter matrix is excessively deleted. When the compression ratio is too small, the potential for deeper com-pression of a matrix is ignored. Considering this problem, we propose a formula to decide optimal compression rate of a parameter matrix. Suppose $g(k)$ repre-sents the compression ratio of a parameter matrix, and $f(k)$ represents the loss rate of a matrix information. Based on Sect. 3.1, the optimal k needs to satisfy

$$\max_{k} \ f(k) + \alpha g(k) \ , \quad where \ 0 \le k \le \frac{mn}{m+n+1}. \tag{5}$$

We use Frobenius norm to measure the retention rate of matrix information:

$$f(k) = \frac{\left\|\widetilde{W}\right\|_{F}}{\|W\|_{F}} \ , \tag{6}$$

and rank compression rate to measure matrix compression rate:

$$g(k) = \frac{r - k}{r} \ . \tag{7}$$

Now (5) becomes to

$$\max_{k} \frac{\left\|\widetilde{W}\right\|_{F}^{2}}{\|W\|_{F}^{2}} + (\frac{r - k}{r})^{2} \ . \tag{8}$$

Based on definition of Frobenius norm,

$$\|W\|_{F} = \sqrt{\sum_{i=1}^{m}\sum_{j=1}^{n} a_{ij}^{2}} = \sqrt{trace(W \times W)} = \sqrt{\sum_{i=1}^{\min(m,n)} \sigma_{i}^{2}} \ , \tag{9}$$

and

$$\left\|\widetilde{W}\right\|_F = \sqrt{\sum_{i=1}^{k} \sigma_i^2} . \tag{10}$$

For the purpose of optimizing this formula more easily, we square $f(k)$ and $g(k)$. So we get

$$\max_k \frac{\sum_{i=1}^{k} \sigma_i^2}{\sum_{i=1}^{\min(m,n)} \sigma_i^2} + \alpha(\frac{r-k}{r})^2 . \tag{11}$$

We set $f(k)$ first-order increment symbol to

$$\nabla f(k) = \frac{\sigma_k^2}{\sum_{i=1}^{\min(m,n)} \sigma_i^2} , \tag{12}$$

first-order increment of $g(k)$

$$\nabla g(k) = \frac{\alpha(2k - 2r - 1)}{r^2} . \tag{13}$$

We can easily see that $f(k)$ is a monotonically increasing function, and the increment is gradually decreasing. $g(k)$ is a monotonically decreasing function and absolute value of the increment is gradually decreasing. Judging from the second-order increment of $f(k)$ and $g(k)$, the change rate of first-order increment of $f(k)$ goes from large to small and the change rate of first-order increment of $g(k)$ is constant. And because front singular values concentrate most of energy, when k is small, the first-order increment of $f(k)$ should be greater than the first-order increment of $g(k)$. Moreover, we assume that k should not reach the optimal value near the upper and lower bounds of the domain. So, when the first-order increment of $f(k)$ is equal to the first-order increment of $g(k)$, the whole formula reaches maximum. So optimal k satisfies

$$\alpha(2r + 1) \sum_{i=1}^{\min(m,n)} \sigma_i^2 = r^2\sigma_k^2 + 2\alpha \sum_{i=1}^{\min(m,n)} \sigma_i^2 k . \tag{14}$$

Among them, α is a hyperparameter, which measures the importance of matrix compression ratio. When the channel is extremely congested and packet loss occurs seriously, clients increase α value appropriately. When network connection is smooth, clients decrease α value. More importantly, it is worth noting that parameters needed for getting k totally depend on singular values of matrix $W \in R_{m \times n}$ mentioned in Sect. 3.1 and do not require additional calculations to occupy computing time of edge nodes. So it does not bring burden to hardware devices and saves cell phone battery power.

3.4 Train Low Rank Parameter Matrix

During truncation, although redundancy and unimportant information are deleted, it is still harmful to neural network powerful expression ability. We hope to minimize the accuracy loss caused by the truncation of singular values. So we train neural network parameters into a rank compact matrix whose energy is more focused on previous singular values. The benefit of the parameter matrix to be low rank for neural network is twofold. First, it brings greater generalization performance. Second, truncation of decomposed matrix sent in the reporting stage will not cause too much loss of accuracy. Generally, we assume neural network loss function l. So our training task is:

$$\min l(x; w) . \tag{15}$$

In order to ensure that the trained matrix is a low-rank matrix, we add the nuclear norm to constrain the parameter matrix W. Now our task is

$$\min l(x; w) + \lambda \|W\|_* . \tag{16}$$

The hyperparameter λ measures the influence of nuclear norm on the entire formula. One strategy is to use subgradient descent to solve this problem. First $W = U\Sigma V$ is SVD of W, then the sub-gradient of the above formula is $\nabla l + UV^T$. The sub-gradient method can be used for non-differentiable objective functions and can be applied to a wider range of problems. However, sub-gradient method is much slower than stochastic gradient descent method, which will cause the problem of slow convergence. Another way is to use proximity operator [26]:

$$W_{k+1} = prox_{\gamma\lambda\|\Delta\|_*}(W_k - \gamma\nabla l(W_k)) \tag{17}$$

to solve the problem. Later, it applies soft-thresholding operator to W. However, it still involves singular value decomposition of the parameter matrix W in each mini batch gradient descent round, which is not time-efficient. So we replace the nuclear norm with Frobrenius norm which has the advantages of being smooth and differentiable to constrain parameter matrix rank. Now our task is:

$$\min l(x; w) + \lambda \|W\|_F . \tag{18}$$

3.5   Noise Reduction Method

Because common machine cannot store gradients of all training samples in its memory, SGD feeds a fraction of samples to models to complete this training round. The direction of the gradient descent completely depends on gradient calculation results of current mini batch, resulting in a large variance. When model gets to around local optimal point, the noise will make it oscillate back and forth around destination and cannot shrink immediately. We treat low-rank truncation of model weight matrix as a kind of noise applied to the gradient. According to Jorge Nocedal et al, there are three ways to reduce variance [27]. The first type is the dynamic sampling method, which reduces the variance of gradient estimation by gradually increasing sample size when calculating gradient. The second type is iterative averaging method which averages the parameter matrix $\widetilde{W}$ obtained after each training round with historically stored parameter matrix to reduce its variance; The third type is gradient aggregation method, which stores historical gradient of each sample and directly or indirectly uses historical or current gradients of all samples to make corrections during each mini-batch estimation of gradient. Compared to the second method, both first and third methods require edge nodes to store many gradients locally. We are more willing to transfer this workload to the parameter server which performs only gradient aggregation task before. Let the server store the historical parameter matrix sent by each client at each round. The method is described as follows. Before server applying weight aggregation, the parameter server first uses the tail averaging method [28]:

$$\widetilde{w_{k+1}} \leftarrow \frac{1}{k-s+1}\Sigma_{j=s+1}^{k+1}\widetilde{w_j} \tag{19}$$

to reduce its variance for each client. Compared to complicated explanations given in [28], it is very easy to understand why (20) only averages the nearest $k-s+1$ historical gradients. According to the definition of model's expected generalization error:

$$E_{model}(f; D) = E_D[(f(x; D) - y)^2] = E_D[(\overline{f}(x) - y)^2] + E_D[(f(x; D) - \overline{f}(x))^2], \tag{20}$$

the initial model has a poor ability to fit the data set, so its variance is large and the bias is small. As the number of training rounds increases, the model becomes more complex, characterized by small bias and large variances. Only recent models can help maintain a low-bias feature while reducing variances.

Table 1. Model

3×3 conv. 96 ReLU
3×3 conv. 96 ReLU
3×3 conv. 96 ReLU stride 2
3×3 conv. 192 ReLU
3×3 conv. 192 ReLU
3×3 conv. 192 ReLU stride 2
3×3 conv. 192 ReLU
1×1 conv. 192 ReLU
1×1 conv. 10 ReLU
Global averaging layer
10 or 100-way softmax

4   Experiment

4.1   Datasets

We evaluate the performance of our algorithm on 3 datasets: MNIST [29], CIFAR-10 and CIFAR-100 [30]. MNIST is a large handwritten digit datasets with a training set of 60,000 examples and a test set of 10,000 examples. The CIFAR-10 data set consists of 10 types of 32×32 color pictures, which contains a total of 60,000 pictures, and each type contains 6000 pictures. 50000 pictures are used as the training set, and 10,000 pictures are used as the test set. The CIFAR-100 dataset has 100 categories, and the number of pictures in each category is one tenth of CIFAR-10.

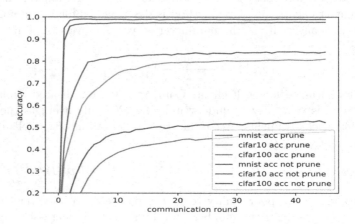

Fig. 4. It shows changes in accuracy on test set and convergence of three datasets after adopting our strategy. Not prune means not adopted.

4.2 Parameter Setting

For us, training a high accuracy model is not our goal. We only care about whether the accuracy can be maintained while hugely improving communication efficiency after using our Fedlr strategy. The learning rate, decay rate, momentum, α, λ is set to 0.01, 10^{-6}, 0.9, 1, 0.01. For Sect. 3.4, sever keeps each client 3 historical gradients. We set 5 clients, each of which gets 10,000 pictures after training dataset has been shuffled. The aggregation algorithm we adopt is Mcmahan federated averaging algorithm [1]. The epoch size of each client training round is 10 and the communication round is 45.

4.3 Model

We use All-CNN-C model (shown in Table 1) in [31] which implementes a convolutional neural network with all convolutional layers. More convolutional layers mean that we have more chances to compress our model, which makes our strategy effects more convincing. This CNN model has a total of eleven layers. The first seven layers are in the form of 4d tensor and can be processed by our strategy. The other layers have much fewer parameters than them. So they are not worth being compressed by our method.

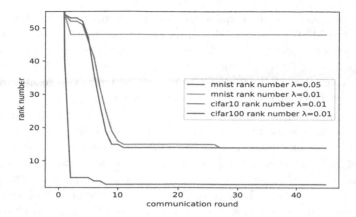

Fig. 5. We show the effect of our strategy by average optimal k drop in the model fifth layer. Initial rank is 192.

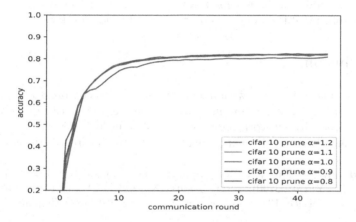

Fig. 6. We show CIFAR-10 datasets test accracy among different α in Sect. 3.2 to see the effect of hyperparameters on the final accuracy of our model.

4.4 Results Analysis

As shown in Fig. 4, the experimental results show that our Fedlr strategy performs very well on three datasets MNIST, CIFAR-10 and CIFAR-100. On MNIST datasets, by sacrificing only 2% accuracy, optimal rank k reduces from 192 to 3 (Fig. 5), achieving 64× parameter matrix rank compression rate. In fact, because the left singular matrix and the right singular matrix are transmitted at the same time, model compression rate is actually only half of rank compression rate, 32×. And model only loses 4% accuracy on CIFAR-10 datasets and convergence rate does not change much compared with normal FL without any compression strategy. Although rank compression rate in the first few rounds is only 4× for three datasets, it improves very fast. Only in a dozen rounds optimal rank k could converge to a fixed value. In hundreds of rounds of FL, these times can be omitted. At the same time, we notice the abnormal performance of rank descent on the MNIST dataset. We think that CNN is very easy to learn the characteristics of digital pictures. So model convergence rate is very fast. The regularization term does not bring constraints to model weights. We need to adjust hyperparameter λ in Sect. 3.3 from 0.01 to 0.05. The result proves our opinion. The CIFAR datasets do not have such problems, so we do not make this adjustment for them. As shown in Fig. 6, We find that final accuracy of the model is not sensitive to the hyperparameter α. We think there are two possible reasons. The first point is that the number of clients is small and owing training set is large, they are very easy to learn core knowledge. The second point is that we train a low-rank parameter matrix, which causes the previous singular values to be large. It is difficult to decrease the optimal k even with a larger hyperparameter α. These two reasons inspire our future work to explore the correlation between the value of α and the accuracy of the model when there are a large number of clients with a small training set. We hope this research will provide a minimum bound for the number of needed filters to reconstruct original convolutional layers and guide the design of low-rank filters.

5 Conclusions

Under FL setting, we use a low-rank communication to compress model parameter matrix. In a communication network, parameter server can quickly receive clients responses without worrying about the obvious loss of accuracy. Our algorithm is verified by extensive experimental evaluation on public datasets.

Acknowledgement. This work was supported by National Key R&D Program of China (No. 2017YFC0803700), NSFC grants (No. 61532021 and 61972155), Shanghai Knowledge Service Platform Project (No. ZF1213) and Zhejiang Lab (No. 2019KB0AB04).

References

1. McMahan, H.B., Moore, E., Ramage, D., Hampson, S.: Communication-efficient learning of deep networks from decentralized data. arXiv preprint arXiv:1602.05629 (2016)
2. Liu, Y., et al.: A communication efficient vertical federated learning framework. arXiv preprint arXiv:1912.11187 (2019)
3. Reisizadeh, A., Mokhtari, A., Hassani, H., Jadbabaie, A., Pedarsani, R.: FedPAQ: a communication-efficient federated learning method with periodic averaging and quantization. arXiv preprint arXiv:1909.13014 (2019)
4. Sattler, F., Wiedemann, S., Müller, K.R., et al.: Robust and communication-efficient federated learning from non-IID data. arXiv preprint arXiv:1903.02891 (2019)
5. Srinivas, S., Subramanya, A., Venkatesh Babu, R.: Training sparse neural networks. In: Proceedings of the IEEE Conference on Computer Vision and Pattern Recognition Workshops, pp. 138–145 (2017)

6. Liu, Z., Li, J., Shen, Z., Huang, G., Yan, S., Zhang, C.: Learning efficient convolutional networks through network slimming. In: Proceedings of the IEEE International Conference on Computer Vision, pp. 2736–2744 (2017)
7. Zhu, M., Gupta, S.: To prune, or not to prune: exploring the efficacy of pruning for model compression. arXiv preprint arXiv:1710.01878 (2017)
8. Frankle, J., Carbin, M.: The lottery ticket hypothesis: finding sparse, trainable neural networks. arXiv preprint arXiv:1803.03635 (2018)
9. Luo, J.H., Wu, J., Lin, W.: Thinet: a filter level pruning method for deep neural network compression. In: Proceedings of the IEEE International Conference on Computer Vision, pp. 5058–5066 (2017)
10. Ullrich, K., Meeds, E., Welling, M.: Soft weight-sharing for neural network compression. arXiv preprint arXiv:1702.04008 (2017)
11. Denton, E.L., Zaremba, W., Bruna, J., LeCun, Y., Fergus, R.: Exploiting linear structure within convolutional networks for efficient evaluation. In: Advances in Neural Information Processing Systems, pp. 1269–1277 (2014)
12. Jaderberg, M., Vedaldi, A., Zisserman, A.: Speeding up convolutional neural networks with low rank expansions. arXiv preprint arXiv:1405.3866 (2014)
13. Guo, Y.: A survey on methods and theories of quantized neural networks. arXiv preprint arXiv:1808.04752 (2018)
14. Shayer, O., Levi, D., Fetaya, E.: Learning discrete weights using the local reparameterization trick. arXiv preprint arXiv:1710.07739 (2017)
15. Han, S., Mao, H., Dally, W.J.: Deep compression: compressing deep neural networks with pruning, trained quantization and huffman coding. arXiv preprint arXiv:1510.00149 (2015)
16. Konečný, J., Richtárik, P.: Randomized distributed mean estimation: accuracy vs. communication. Front. Appl. Math. Stat. **4**, 62 (2018)
17. Alistarh, D., Grubic, D., Li, J., Tomioka, R., Vojnovic, M.: QSGD: communication-efficient SGD via gradient quantization and encoding. In: Advances in Neural Information Processing Systems, pp. 1709–1720 (2017)
18. Horvath, S., Ho, C.Y., Horvath, L., Sahu, A.N., Canini, M., Richtarik, P.: Natural compression for distributed deep learning. arXiv preprint arXiv:1905.10988 (2019)
19. Wu, J., Huang, W., Huang, J., Zhang, T.: Error compensated quantized SGD and its applications to large-scale distributed optimization. arXiv preprint arXiv:1806.08054 (2018)
20. Suresh, A.T., Yu, F.X., Kumar, S., McMahan, H.B.: Distributed mean estimation with limited communication. In: Proceedings of the 34th International Conference on Machine Learning, vol. 70, pp. 3329–3337. JMLR. org (2017)
21. Caldas, S., Konečny, J., McMahan, H.B., Talwalkar, A.: Expanding the reach of federated learning by reducing client resource requirements. arXiv preprint arXiv:1812.07210 (2018)
22. Prabhavalkar, R., Alsharif, O., Bruguier, A., McGraw, L.: On the compression of recurrent neural networks with an application to LVCSR acoustic modeling for embedded speech recognition. In: 2016 IEEE International Conference on Acoustics, Speech and Signal Processing (ICASSP), pp. 5970–5974. IEEE (2016)
23. Langeberg, P., Balda, E.R., Behboodi, A., Mathar, R.: On the effect of low-rank weights on adversarial robustness of neural networks. arXiv preprint arXiv:1901.10371 (2019)
24. Kalman, D.: A singularly valuable decomposition: the SVD of a matrix. Coll. Math. J. **27**(1), 2–23 (1996)

25. Koneçný, J., McMahan, H.B., Yu, F.X., Richtárik, P., Suresh, A.T., Bacon, D.: Federated learning: strategies for improving communication efficiency. arXiv preprint arXiv:1610.05492 (2016)
26. Ciliberto, C., Stamos, D., Pontil, M.: Reexamining low rank matrix factorization for trace norm regularization. arXiv preprint arXiv:1706.08934 (2017)
27. Bottou, L., Curtis, F.E., Nocedal, J.: Optimization methods for large-scale machine learning. SIAM Rev. **60**(2), 223–311 (2018)
28. Jain, P., Kakade, S.M., Kidambi, R., Netrapalli, P., Sidford, A.: Parallelizing stochastic approximation through mini-batching and tail-averaging. STAT **1050**, 12 (2016)
29. LeCun, Y., Bottou, L., Bengio, Y., Haffner, P.: Gradient-based learning applied to document recognition. Proc. IEEE **86**(11), 2278–2324 (1998)
30. Krizhevsky, A., Hinton, G.: Learning multiple layers of features from tiny images (2009)
31. Springenberg, J.T., Dosovitskiy, A., Brox, T., Riedmiller, M.: Striving for simplicity: the all convolutional net. arXiv preprint arXiv:1412.6806 (2014)

Question Answering over Knowledge Base with Symmetric Complementary Attention

Yingjiao Wu[✉] and Xiaofeng He[✉]

School of Software Engineering, East China Normal University, Shanghai, China
51174500052@stu.ecnu.edu.cn, hexf@cs.ecnu.edu.cn

Abstract. Knowledge Base Question Answering (KBQA), which aims to answer natural language questions with structured data from a knowledge base is an important Natural Language Processing (NLP) problem. To answer the question, we need to find the fact from the Knowledge Base whose subject and relation best match the question. Most existing methods treat this task as a pipeline of two separate subtasks: subject matching and relation matching. While ignoring the relevance between them. In this paper, we focus on solving this problem through a joint learning method. We present a neural joint model with a shared encoding layer to learn the two subtasks together to improve each other. In particular, we design a Symmetric Bidirectional Complementary Attention module based on the attention mechanism and the gate mechanism to model the relationship between the two subtasks. The experimental results demonstrate that our approach can obtain higher accuracy than the state-of-the-art method.

Keywords: Question answering · Knowledge base · Joint learning

1 Introduction

Knowledge Base Question Answering, which aims to automatically answer natural language questions with structured data from a knowledge base is an important NLP problem recently. Since large-scale knowledge bases (KBs), such as DBpedia [13] and Freebase [3] are freely available, KBQA has attracted much attention in QA research. The KB usually contains a large set of structured representation of facts in the form of the triple (*subject, relation, object*). KBQA provides an efficient way to access these valuable resources. It has widely used in search engines, chatting robots, customer service and so on.

A Simple KBQA task involves answering a question such as *"who is the producer of the empire strikes first?"* which asks for a single fact (*the empire strikes first, music.release.producers, brett gurewitz*). It is the most common type in KBQA tasks and can be considered as the basic component of complex question. To find the answer to such questions, there are two main steps: (1) candidate

Y. Nah et al. (Eds.): DASFAA 2020 Workshops, LNCS 12115, pp. 17–31, 2020.
https://doi.org/10.1007/978-3-030-59413-8_2

generation: to retrieve candidate facts related to the question from the knowledge base. (2) candidate reranking: to rerank candidate facts according to the their similarity with the question.

For candidate generation, we aim to narrow the search space from the entire knowledge base to a question-related subgraph. There are two steps, the first step is entity mention detection, which aims to identify the boundaries of named entity in the question. The next step is entity linking, which aims to link entity mention with their corresponding subjects in KB [23]. The most common solution to this task is to retrieve subjects from KB whose name or alias match an n-gram of entity mention. Finally, the facts related to these subjects are selected as a candidate answer.

For candidate reranking, existing studies typically learn the semantic representations of questions and candidate facts so that representations of questions and their gold facts are close to each other in the semantic space [4]. Many previous work formulate it as a multi-stage task, in each stage, one element of the fact is compared with the question to produce a partial similarity score by a dedicated model, then these partial scores are combined to generate the overall score [7]. The candidate fact with the highest score will be selected as the correct fact. For example, for the question *"who is the producer of the empire strikes first?"* and a candidate fact (*the empire strikes first,music.release.producers,brett gurewitz*), they construct models to match the entity mention *"the empire strikes"* in question with candidate subjects named *"the empire strikes first"* in KB and match the question pattern *"who is the producer of X?"* (replace the entity mention in the question with a special token X) with relation named *"music.release.producers"*. They are referred to as mention-subject matching and pattern-relation matching respectively. Despite the effective as these existing approach, there still remains some challenges among these stages:

1) *Propagation of errors*: in the candidate generation step, incorrect entity mention detection will result in incorrect entity linking. In the candidate reranking step, the mismatch of mention-subject may lead to the mismatch of pattern-relation.
2) *Ambiguity of entities*: a named entity may have multiple surface forms, such as its full name, partial name, alias. Besides, an entity name may correspond to many subject nodes in KB. For example, there are dozens of subjects named *"collection"*.
3) *Out of vocabulary*: due to the large scale of the knowledge base, the training data usually only cover a part of the facts in the KB. However, the trained model may perform poorly when answering the questions corresponding with unseen words especially unseen relation in the training process.

Faced with the above problems, we construct a framework for the KBQA task as illustrated in Fig. 1. For candidate generation, we use a heuristic matching method based on various matching methods for entity linking, indirectly modifying the errors of entity mention in the question. For candidate reranking, we propose a neural joint model to further alleviate error propagation by jointly scoring the subjects and relations of candidate facts. Our model not only

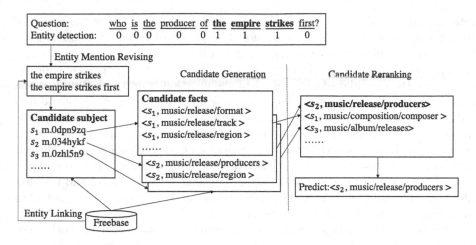

Fig. 1. The overview of simple KBQA, for a given question *"who is the producer of the empire strikes first"*, we first identify the entity mention in the question by the sequence labeling model. During the entity linking step, we use a heuristic matching method to retrieve candidate subjects. The candidate set is all the facts related to those subjects. Then rerank the candidate facts through a joint scoring model, the answer to the question is the object of the top-1 fact.

considers the relevance of the subject matching and the relation matching but also incorporates more information for semantic matching. All the parameters could be jointly trained through backpropagation. Furthermore, we use a CNN-based character encoder so that every token's vector can be formed even it is out-of-vocabulary words. The experimental results demonstrate that our model is better than the state-of-the-art method.

The rest of this paper is organized as follows: Sect. 2 discusses related research. Section 3 declares the details of the proposed method. Section 4 describes the dataset and experiments. We analyze the results in Sect. 4.3 and conclude our work in Sect. 5.

2   Related Work

There are two mainstream research lines for the KBQA task. One research line is based on semantic parsing [2,5,9,21,24,27]. The main idea is to transform natural language questions into logical forms. The traditional semantic parsing method is the symbol-based method which usually gets the logical form of the question by the logical language Lambda Dependency based Compositional Semantics (λ - DCS) [1,2] and gets the semantic representations from the logic forms by Combinatory Categorical Grammars (CCG) [12,21,30]. Recently, neural-symbolic based semantic parsing methods have emerged with the development of deep learning. The main idea is to use an encoder-decoder model to encode the question into a context vector, and then decode the context vector into the logical form [6,16,25]. There are various models, such as the sequence to sequence model, sequence to tree model, sequence to action model. However, this method requires complex annotation and it is difficult to scale.

The other research line is based on candidate reranking, the main idea is to retrieve the candidate facts related to the question from the KB and rerank them with a match-scoring model. For candidate generation, many of the previous works use a sequence labeling model to detect the entity mention in the question and then use all the n-grams of entity mention to match the subject nodes in KB[7,28]. Candidate facts are generated from the subgraphs of the KB concerning only involved subjects [26]. For candidate reranking, the core idea is to build a model to learn semantic representations of the question and the candidates, such that the correct supporting evidence will be the nearest neighbor of the question in the learned vector space [7].

Recently, with the progress of deep learning, various neural network models are used for representation learning of questions and candidate facts. [8] focused on answer selection, they use three columns of CNNs to represent the question with three different answer aspects, i.e. answer path, answer context and answer type. [7] formulate a two-step probabilistic treatment of the KBQA problem using BiGRU network and word embedding, which modeling $p(r|q)$ to infer the target relation of the question, and then modeling $p(s|q, r)$ to get the target subject from the candidate subjects related to the best matching relation. [28] focused on joint fact selection, they use a character-based CNN model for mention-subject matching and use a word-based CNN with attentive max pooling for pattern-relation matching. [17] implemented an end-to-end model by leveraging a word-level and character-level question encoder with the GRU model. [29] improve the performance of relation matching by using deep hierarchical residual Bi-LSTM to match the question and candidate relations via two different granularity, i.e. relation-level (present the relation as a single token) and word-level (split relation name into word sequence). [10] introduced a pattern revising method during pattern extraction and entity linking stage, and then multi-level encodings and multi-dimension information are used in the joint fact selection stage. Our work is based on the candidate reranking method, and following the method proposed by [7], we also employ a probabilistic treatment of the KBQA task.

In this work, we study how to utilize joint learning to help knowledge-based questions answering. Joint learning originates from multi-task learning, which aims to improve the performance of all the tasks by learning the shared information between the tasks. Recently, joint learning has achieved significant improvements in information extraction tasks [18,31]. Most of the exsist works use a shared encoding layer to extract shared knowledge between two subtasks: entity extraction and relation extraction. Inspired by this, we also use a neural joint model with a shared encoding layer to jointly score the subject and relation of a candidate fact.

3   Overview

In this section, we will introduce our method in detail. For a given a natural language question q, let $\mathcal{K} = \{(s_i, r_i, o_i), i = 1, 2, ..., |\mathcal{K}|\}$ be a background knowledge base. Where s_i, r_i and o_i represent the subject, relation and object of a fact respectively. And $|\mathcal{K}|$ represents the number of all facts in the KB, it is usually tens of millions. The task of KBQA is to find the best match fact in the knowledge base $\mathcal{K}$. Obviously, we only need to consider the facts related to the question as a candidate set $\mathcal{C}(\mathcal{C} \in \mathcal{K})$. The problem can be formulated into a probabilistic form [7], which to find the subject-relation pair $\hat{s}, \hat{r}$ from candidate set $\mathcal{C}$ which maximizes the joint conditional probability $p(s, r|q)$:

$$\hat{s}, \hat{r} = \arg \max_{s, r \in \mathcal{C}} p(s, r|q) \tag{1}$$

So solving this problem will go through two steps: candidate generation and candidate reranking. As shown in Fig. 1, for the candidate generation step, we first detect entity mention in the question that describes the subject of the fact, and then construct a set of candidate facts so that the subject of each fact is associated with the question. For candidate rerank, we estimate the correctness of every candidate fact given the question q and then select the best one. There are two ways to decompose the joint conditional probability $p(s, r|q)$:

$$p(s, r|q) = p(r|q) * p(s|q, r) = p(s|q) * p(r|q, s) \tag{2}$$

The original function can also be expressed as the following formula:

$$\log p(s, r|q) = \log p(r|q) + \log p(s|q, r) = \log p(s|q) + \log p(r|q, s) \tag{3}$$

Then the above conditional probability can be parameterized by the neural network. Most existing methods treat this task as a pipeline of two separate tasks, such as modeling $p(r|q)$ with a relation network and modeling $p(s|q, r)$ with a subject network. Although the pipeline method can be more flexible to design the system, it neglects the relevance of subtasks and may also lead to error propagation [15]. To avoid this problem, we change the factorization as follows:

$$\log p(s, r|q) = \frac{1}{2}[\log p(r|q) + \log p(s|q, r) + \log p(s|q) + \log p(r|q, s)] \tag{4}$$

We design a joint model that contains the subject matching module $[\log p(s|q) + \log p(s|q, r)]$ and the relation matching module $[\log p(r|q) + \log p(r|q, s)]$ in a unified architecture. The question and candidate facts share the same Bidirectional Long Short-term Memory (Bi-LSTM) encoding layer to make two subtasks not independent. In particular, we design a symmetric bidirectional complementary attention module to capture the dependencies between the two subtasks.

The training objective of our model is to jointly learn the parameters of the subject matching module and the relation matching module by maximizes the joint conditional probability:

$$\mathcal{L} = \arg \max_{s, r \in \mathcal{K}} [\log p(r|q) + \log p(s|q, r) + \log p(s|q) + \log p(r|q, s)] \tag{5}$$

3.1   Candidate Generation

Given a question, we first detect entity mention m in the question, which is also a substring of the question. Based on previous works, the entity mention detection is formulated as a sequence labeling problem. We label each token of the question with 0 or 1 to indicate either ENTITY or NOTENTITY through a BiLTSM model [19]. And then retrieve corresponding subject from knowledge base $\mathcal{K}$ whose name matches the entity mention exactly.

However, the wrong entity mention may lead to wrong subject retrieval, which will cause error propagation. There are two main errors of entity mention, the spelling errors in the question and the prediction errors in the sequence labeling model. To relieve this problem, on the one hand, we use fuzzy matching method to alleviate the spelling errors. We measure the Jaccard distance[1] between the token set of entity mention and the token set of candidate subject's name, then filter out subjects whose Jaccard Distances are below the thresholds. On the other hand, we use an extend matching method to modify the prediction errors. To be more specific, expand or narrow at most 2 tokens around m in the question to get new entity mention m', and then use it to retrieve corresponding subjects from $\mathcal{K}$. If candidate subject set $\mathcal{S}$ is still empty, use all the n-grams of the question to match the subjects in $\mathcal{K}$. Finally, candidate facts $\mathcal{C}$ is defined as:

$$\mathcal{C} = \{(s, r, o) : s \in \mathcal{S}, s \to r\} \tag{6}$$

where $s \to r$ represents the relation r connected to the subject s in the $\mathcal{K}$.

3.2   Joint Scoring Model

The joint model tries to match the question with the corresponding candidate facts. Figure 2 illustrates the overview of the model. The input token is represented as a concatenation of character level encoding and word embedding. Then use a Bi-LSTM layer to get hidden representation for each token by taking into account the context on both sides of the token. Unlike the traditional method for candidate fact matching that the elements of the candidate facts are matched with the question separately. Here we capture the relevance between subject matching and relation matching by a symmetric complementary attention (SY-CM-ATTN) module. It makes the representations of the questions in two subtasks are complementary to each other. Both subtasks are matched through a shared convolutional neural network (CNN) layer. Finally, the matching results of the two subtasks are integrated through a linear layer to get the final score.

Character-Level Encoder. The trained model may perform poorly when dealing with the words not seen during training. This problem could be considered as an out of vocabulary (OOV) problem. In recent years, character features have been demonstrated to be effective for the OOV problem. For example, words prefixed with "mis" (such as *mistake, mislead*) are more likely to mean wrong.

---

[1] $\frac{M \cap N}{M \cup N}$, where M, N is the token set of *entity mention* and *subject name* respectively.

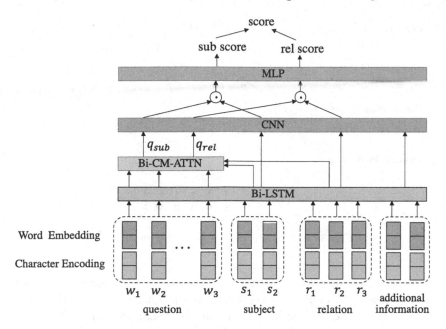

Fig. 2. The overview of the Neural Joint model.

CNN is good at extracting the morphological information of the words. Following previous work [14] we use character embedding followed by a CNN layer to encode each token into a character-level vector representation. As shown in Fig. 3, the character embedding is similar to the word embedding. Both learn an embedding matrix so that the one-hot representation of the character can be mapped to a low-dimensional vector representation. The architecture of the CNN layer is almost the same as that proposed by [11]. In our work, we apply several kernels with size 1 to 3 on the character embedding matrix of the inputs, and then use max-pooling to all filter outputs. So the joint model takes the concatenation of the character encoding and word embedding of each token as inputs, given by:

$$e^{w_i} = [w_i^{char}; w_i^{word}] \tag{7}$$

where $[;]$ denotes the vector concatenation operation. Therefore, each input token w_i is encoded as a fixed-length vector $e^{w_i} \in \mathbb{R}^{H_e \times 1}$, where H_e is the sum of the dimensions of the token's character encoding w_i^{char} and the token's word embedding w_i^{word}. And e^{w_i} is further encoded by the Bi-LSTM layer to produce hidden vector $h_i \in \mathbb{R}^{H_d}$, where H_d is the dimension of the hidden state.

Symmetric Complementary Attention. The question is mainly concerned with the subject and relation of the fact. However, we find that the information distributions of subject and relation in the question are complementary. For example, for the question *"who is the producer of the empire strikes*

Concatenation

Max-pooling Layer

Convolutional Layer

Character Embedding

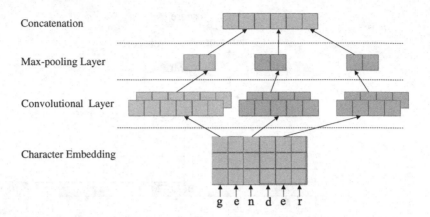

g e n d e r

Fig. 3. The architecture of character encoder.

first?", the subject named *"the empire strikes first"* will pay more attention to those tokens *"the empire strikes first"* in the question, while the relation *"music.release.producers"* tends to focus on the remaining tokens *"who is the producer of"*. So we model the relationships by designing a symmetric complementary attention module in the joint model. As shown in Fig. 4, we first get the attention distribution of the subject and the relation to the question respectively. Then get corresponding complementary attention through output gate. These two types of attention can work together in each module for better semantic representation. For simplicity, the attention mechanism is not shown explicitly in the figure.

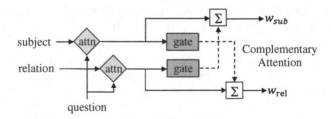

Fig. 4. The detail of the SY-CM-ATTN module.

For the calculation of the attention distribution. We follow the work on answer selection [22]. Given $Q \in \mathbb{R}^{H_d \times M}$, $S \in \mathbb{R}^{H_d \times N}$ and $R \in \mathbb{R}^{H_d \times L}$ as the encoding results of the question, subject and relation respectively, where M, N, L represent the length of the question, subject and relation respectively. The alignment scoresbetween the tokens in question and tokens in subject as follows:

$$A_{sub} = \tanh(Q^T W_a S) \tag{8}$$

where $W_a \in \mathbb{R}^{H_d \times H_d}$ is a matrix of parameters to be learned by the network. We apply row-wise max poolings over A_{sub} to generate the vectors $a \in \mathbb{R}^{M \times 1}$. Each element of the vector a indicates how important the token in the question is to the subject. The vector is then normalized by the softmax function as follow:

$$[\alpha]_i = \frac{\exp [a]_i}{\sum_{j=0}^{M} \exp [a]_j} \tag{9}$$

it makes the values in the vector to sum up to 1 and each individual value will lie between 0 and 1, therefore get the attention weight vector of subject to the question $\alpha \in \mathbb{R}^{M \times 1}$. In the same way, we can also get the alignment scores A_{rel} between the question and the relation, the weight vector $\beta \in \mathbb{R}^{M \times 1}$ of the relation to the question as follows:

$$A_{rel} = \tanh(Q^T W_b R) \tag{10}$$

$$b = Row\_Maxpooling(A_{rel}) \tag{11}$$

$$[\beta]_i = \frac{\exp [b]_i}{\sum_{j=0}^{M} \exp [b]_j} \tag{12}$$

For the calculation of the complementary attention distribution. Inspired by the gate mechanism in LSTM, we design the output gate f_o to control how much attention information can flow from one module to another. Specifically, the complementary attention weight of the relation to the i-th token in question calculates as:

$$[\bar{\beta}]_i = f_o([\beta]_i) \tag{13}$$

$$f_o([\beta]_i) = 1 - [\beta]_i \tag{14}$$

The complementary attention of the relation to question can be helpful to the match of subject and question, so we combine the information to get the overall attention weight of the subject to the i-th token in question as follows:

$$[wt_{sub}]_i = \lambda * [\alpha]_i + (1 - \lambda) * [\bar{\beta}]_i \tag{15}$$

Where λ is a hyperparameter to control the impact of complementary attention weights. The first part to the right of the above equation focuses on modeling $p(s|q)$, while the complementary attention on the second part focuses on modeling $p(s|q, r)$ by taking into account the information of relation. In the same way, we can get the complementary attention of subject to the i-th token of the question, and the overall attention of the relation to the i-th token of the question as follows:

$$[\bar{\alpha}]_i = f_o([\alpha]_i) \tag{16}$$

$$[wt_{rel}]_i = \lambda * [\beta]_i + (1 - \lambda) * [\bar{\alpha}]_i \tag{17}$$

The two parts to the right of the equation modeling $p(r|q)$ and $p(r|q, s)$ respectively. Therefore we get the overall attention distribution of subject to question

$wt_{sub} \in \mathbb{R}^{M \times 1}$ and relation to question $wt_{rel} \in \mathbb{R}^{M \times 1}$. Finally, the representation of question q_{sub} and q_{rel} are calculated by:

$$q_{sub} = Q \odot wt_{sub} \tag{18}$$

$$q_{rel} = Q \odot wt_{rel} \tag{19}$$

where $\odot$ is the element-wise product of vectors, $q_{sub} \in \mathbb{R}^{H_d \times M}$, $q_{rel} \in \mathbb{R}^{H_d \times M}$.

Semantic Matching. After getting the different representations of the question for subject and relation, a shared CNN layer is used to encode inputs as a fixed-length vector, so that we can measure the similarity in the same semantic space. The structure of CNN is the same as that used in the character-level encoder. We use the element-wise product followed by MLP to measure the similarity between two vector representations. Then the scores of the two modules are combined through a linear layer to get the overall score of candidate fact. During this process, relation matching is the key and difficult point in the KBQA task. Because the relation name is short and fixed whereas the corresponding pattern in the question is highly variable in length and word choice [28]. Hence we combine information of the subject type and the object type to enhance the representation of the relation, which can also help for subject disambiguation. Finally, we rerank the candidate facts based on the scores of the joint model, and select the best one as the answer to the question.

Training. The joint model is trained by the hinge loss which tries to separate the score of each negative fact from the positive fact by a margin:

$$\mathcal{L} = \sum \max(0, \gamma - s(q, [s, r]^+) + s(q, [s, r]^-)) \tag{20}$$

where γ is the margin; $[s, r]^+$ is the ground true subject and relation for the question q, $[s, r]^+$ is the incorrect subject and relation for the question q, $s(\cdot, \cdot)$ represents the scoring result of the joint model.

4   Experiments

4.1   Dataset and Evaluation

We train and evaluate our model on **SimpleQuestion** benchmark [4] which provides 108,441 questions labeled with a single Freebase triple and two subsets of Freebase: FB2M (2M entities, 0.006M relations, 14M atomic fact triples), FB5M (5M entities, 0.007M relations, 22M atomic fact triples). The statistics of the dataset are shown in Table 1 and Table 2.

The evaluation metric is path-level accuracy which takes the prediction as correct if the subject and the relation are correctly retrieved.

$$Accuracy = \frac{\sum_{i=1}^{N} \mathbb{1}[(\hat{s}_i, \hat{r}_i) = (s_i, r_i)]}{N} \tag{21}$$

Table 1. Statistics of the dataset.

Dataset	#Train	#Dev	#Test
SimpleQuestion	75,910	10,845	21,687

Table 2. Statistics of the related Knowledge Base.

KB	#Entities	#Relations	#Atomic Facts
FB2M	2,150,604	6,701	14,180,937
FB5M	4,904,397	7,523	22,441,880

4.2 Experimental Settings

During training, all word embeddings are initialized using the pre-trained GloVe [20] with 300 dimensions. For the CNNs, we use multiple filters with widths 1, 2 and 3, the number of filters for each size is 100, in this way, the dimension of the character encoding vector is the same as the dimension of the word embedding (300). After contacting the character encoding and the word embedding, we get an input vector with a dimension of 600 for each token. The hidden layer of Bi-LSTM has size 300 so that the dimension of Bi-LSTM encoding result is equal to the dimension of the input vector (600). The dropout rate for the CNNs was 0.5 and 0.3 for the RNNs. The margin for the hinge loss γ is set to 0.1. The optimizer used for training the models is Adam and the initial learning rate is 0.0001. The negative sample number is 50 for the joint model, they are mainly generated from the samples that are literally similar to positive samples in the candidate set. The hyperparameter λ in the network is 0.5. All hyperparameters are determined according to the performance on the validation set.

4.3 Results

Table 3. Comparison with baseline methods on SimpleQuestions benchmark.

Approach	Accuracy-RC	Accuracy
AMPCNN [28]	91.3	76.4
HR-BiLSTM [29]	93.3	77.0
Pattern-revising [10]	–	80.2
SY-CM ATTN (Our method)	**93.7**	**80.5**

To demonstrate the effectiveness of the proposed approach, we compare our method with several recent baselines. Table 3 shows the results on SimpleQuestions dataset (base on FB2M knowledge base). The third column represents the accuracy of the answer prediction. AMPCNN method and HR-BiLSTM method aim to enhance the performance of KBQA by designing better relation matching models. AMPCNN method proposes a joint model that matches the subject by character-based CNN and matches the relation by word-based CNN followed by the attentive max-pooling respectively. HR-BiLSTM method proposes a hierarchical Bi-LSTM with residual connection to match question and relation from two different granularity, i.e. relation-level and word-level. The Pattern-revising method in [10] adopts a two-stage framework that first selects top-n candidate subjects by a relation detection model and then reranks candidate facts by the joint model, the joint model is adapted based on the model in AMPCNN, and yield the state-of-the-art results.

As shown in Table 3, our approach achieves an accuracy of 80.5%, which is better than the state-of-the-art results by 0.3 points. The improvement is marginal, however, our method jointly scores the candidate facts in a single step compared with the state-of-the-art method. It also proves that joint learning is promising in simple KBQA tasks.

In addition, we also explore the effectiveness of relation classification (RC) subtask, RC task is created based on SimpleQuestion: label each question with the ground true subject and relation, all the other relations of the gold subject are labeled as negative [28]. The second column in Table 3 represents the results on the RC subtask. Both AMPCNN method and HR-BiLSTM method convert the question into a question pattern (by replacing the true entity mention in the question as special characters such as X), and then model the match between the question pattern and relation. In our joint model, we directly calculate the similarity between the subject, relation and the question. As a result, our approach improves the RC subtask by 0.4 points and achieves the state-of-the-art result. It proves that we can improve the performance of subtasks through the joint learning method.

4.4 Ablation Study

Table 4. The ablation results of our approach.

Approach	Setting	Accuracy
Our approach	fully integrated	**80.5**
Our approach	w/o rel CM ATTN	80.1
Our approach	w/o sub CM ATTN	79.8
Our approach	w/o both	79.1

In this part, we further analyze the effectiveness of different parts of the framework. As shown in Table 4, the overall accuracy of our joint model is 80.5%. When removing the complementary attention of the subject or the relation, the overall accuracy of the model decrease at least 0.4, indicating that the complementary attention information of either module is important for our framework. It is interesting to find that the results shown in the third line perform worse than the second line by 0.3 points, showing that the complementary information of the subject is more important to our framework. The performance decreases obviously when we remove any of them, which demonstrates that the performance of the joint model can be enhanced by the symmetric bidirectional complementary module.

4.5 Error Analysis

We perform error analysis on our approach by randomly sampling 100 wrong answered questions. There are three main types of errors.

The most common kind of error (52%) is that the true subject is in the candidate set, but has not been predicted correctly. The main reason is that there are many subjects with the same name and similar context in the knowledge base. And the knowledge base or question can't support more evidence to distinguish them. Another kind of error case (43%) is that the true subject is not selected in the candidate set. Since the training data is obtained by human annotators to generate questions from the selected facts, spelling errors in the question are unavoidable, which makes the entity mention in the question can't be detected correctly. The last type of error (5%) is the relation prediction error, that is, the subject is predicted correctly, but a wrong relation is chosen. For example, for similar candidate relations *"release.tracks"* and *"release.track_list"*, both correspond to the question pattern *"what's a track from X"*. From the error statistics, we can find that subject selection is more essential than relation selection, it may be the major bottleneck of Simple KBQA.

5 Conclusion

In this paper, we study how to utilize joint learning to help knowledge based question answering. Specifically, we propose a joint model that aims to alleviate the problem of error propagation in the KBQA. In our model, a character encoder is used to mitigate the OOV problems. And we design a novel symmetric bidirectional complementary attention module to model the dependency between two subtasks. In addition, the various contextual information of candidate facts is used to improve the performance of semantic matching. The parameters of the model can be jointly learned during the training process. Empirical results show that our model improves the performance of KBQA.

References

1. Lambda dependency-based compositional semantics. CoRR, abs/1309.4408 (2013)
2. Berant, J., Chou, A., Frostig, R., Liang, P.: Semantic parsing on freebase from question-answer pairs. In: Proceedings of the 2013 Conference on Empirical Methods in Natural Language Processing, EMNLP, pp. 1533–1544 (2013)

3. Bollacker, K., Evans, C., Paritosh, P., Sturge, T., Taylor, J.: Freebase: a collaboratively created graph database for structuring human knowledge. In: Proceedings of the 2008 ACM SIGMOD International Conference on Management of Data, SIGMOD, pp. 1247–1250 (2008)
4. Bordes, A., Usunier, N., Chopra, S., Weston, J.: Large-scale simple question answering with memory network. CoRR, abs/1506.02075 (2015)
5. Cai, Q., Yates, A.: Large-scale semantic parsing via schema matching and lexicon extension. In: Proceedings of the 51st Annual Meeting of the Association for Computational Linguistics (Volume 1: Long Papers), pp. 423–433. ACL (2013)
6. Chen, B., Sun, L., Han, X.: Sequence-to-action: end-to-end semantic graph generation for semantic parsing. In: Proceedings of the 56th Annual Meeting of the Association for Computational Linguistics (Volume 1: Long Papers), pp. 766–777. ACL (2018)
7. Dai, Z., Li, L., Xu, W.: CFO: conditional focused neural question answering with large-scale knowledge bases. In: Proceedings of the 54th Annual Meeting of the Association for Computational Linguistics (Volume 1: Long Papers). ACL (2016)
8. Dong, L., Wei, F., Zhou, M., Xu, K.: Question answering over freebase with multi-column convolutional neural networks. In: Proceedings of the 53rd Annual Meeting of the Association for Computational Linguistics and the 7th International Joint Conference on Natural Language Processing (Volume 1: Long Papers), pp. 260–269. ACL (2015)
9. Fader, A., Zettlemoyer, L., Etzioni, O.: Open question answering over curated and extracted knowledge bases. In: Proceedings of the 20th ACM SIGKDD International Conference on Knowledge Discovery and Data Mining, KDD, pp. 1156–1165 (2014)
10. Hao, Y., Liu, H., He, S., Liu, K., Zhao, J.: Pattern-revising enhanced simple question answering over knowledge bases. In: Proceedings of the 27th International Conference on Computational Linguistics, COLING, pp. 3272–3282 (2018)
11. Kim, Y.: Convolutional neural networks for sentence classification. In: Proceedings of the 2014 Conference on Empirical Methods in Natural Language Processing, EMNLP, pp. 1746–1751 (2014)
12. Kwiatkowski, T., Zettlemoyer, L., Goldwater, S., Steedman, M.: Lexical generalization in CCG grammar induction for semantic parsing. In: Proceedings of the 2011 Conference on Empirical Methods in Natural Language Processing, EMNLP, pp. 1512–1523 (2011)
13. Lehmann, J., et al.: DBpedia - a large-scale, multilingual knowledge base extracted from Wikipedia. Semant. Web **6**(2), 167–195 (2015)
14. Li, F., Zhang, M., Guohong, F., Ji, D.: A neural joint model for entity and relation extraction from biomedical text. BMC Bioinform. **18**(1), 198:1–198:11 (2017)
15. Li, Q., Ji, H.: Incremental joint extraction of entity mentions and relations. In: Proceedings of the 52nd Annual Meeting of the Association for Computational Linguistics (Volume 1: Long Papers), pp. 402–412. ACL (2014)
16. Liang, C., Berant, J., Le, Q., Forbus, K.D., Lao, N.: Neural symbolic machines: learning semantic parsers on freebase with weak supervision. In: Proceedings of the 55th Annual Meeting of the Association for Computational Linguistics, pp. 23–33. ACL (2017)
17. Lukovnikov, D., Fischer, A., Lehmann, J., Auer, S.: Neural network-based question answering over knowledge graphs on word and character level. In: Proceedings of the 26th International Conference on World Wide Web, WWW, pp. 1211–1220 (2017)

18. Miwa, M., Bansal, M.: End-to-end relation extraction using LSTMs on sequences and tree structures. In: Proceedings of the 54th Annual Meeting of the Association for Computational Linguistics (Volume 1: Long Papers). ACL, The Association for Computer Linguistics (2016)

19. Mohammed, S., Shi, P., Lin, J.: Strong baselines for simple question answering over knowledge graphs with and without neural networks. In: Proceedings of the 2018 Conference of the North American Chapter of the Association for Computational Linguistics: Human Language Technologies (Volume 2: Short Papers), NAACL-HLT, pp. 291–296 (2018)

20. Pennington, J., Socher, R., Manning, C.: Glove: global vectors for word representation. In: Proceedings of the 2014 Conference on Empirical Methods in Natural Language Processing, EMNLP, pp. 1532–1543 (2014)

21. Reddy, S., Lapata, M., Steedman, M.: Large-scale semantic parsing without question-answer pairs. Trans. Assoc. Comput. Linguist. **2**, 377–392 (2014)

22. dos Santos, C., Tan, M., Xiang, B., Zhou, B.: Attentive pooling networks. CoRR, abs/1602.03609 (2016)

23. Shen, W., Wang, J., Han, J.: Entity linking with a knowledge base: issues, techniques, and solutions. IEEE Trans. Knowl. Data Eng. **27**(2), 443–460 (2015)

24. Xu, K., Reddy, S., Feng, Y., Huang, S., Zhao, D.: Question answering on freebase via relation extraction and textual evidence. In: Proceedings of the 54th Annual Meeting of the Association for Computational Linguistics (Volume 1: Long Papers). ACL (2016)

25. Xu, K., Wu, L., Wang, Z., Yu, M., Chen, L., Sheinin, V.: Exploiting rich syntactic information for semantic parsing with graph-to-sequence model. In: Proceedings of the 2018 Conference on Empirical Methods in Natural Language Processing, EMNLP, pp. 918–924 (2018)

26. Yao, X., Van Durme, B.: Information extraction over structured data: question answering with freebase. In: Proceedings of the 52nd Annual Meeting of the Association for Computational Linguistics (Volume 1: Long Papers), pp. 956–966. ACL (2014)

27. Yih, S.W., Chang, M.-W., He, X., Gao, J.: Semantic parsing via staged query graph generation: question answering with knowledge base. In: Proceedings of the 53rd Annual Meeting of the Association for Computational Linguistics (Volume 1: Long Papers), pp. 1321–1331. ACL (2015)

28. Yin, W., Yu, M., Xiang, B., Zhou, B., Schütze, H.: Simple question answering by attentive convolutional neural network. In: 26th International Conference on Computational Linguistics, COLING, pp. 1746–1756 (2016)

29. Yu, M., Yin, W., Hasan, K.S., dos Santos, C., Xiang, B., Zhou, B.: Improved neural relation detection for knowledge base question answering, pp. 571–581 (2017)

30. Zettlemoyer, L., Collins, M.: Online learning of relaxed CCG grammars for parsing to logical form. In: Proceedings of the 2007 Joint Conference on Empirical Methods in Natural Language Processing and Computational Natural Language Learning, EMNLP-CoNLL, pp. 678–687 (2007)

31. Zheng, S., et al.: Joint entity and relation extraction based on a hybrid neural network. Neurocomputing **257**, 59–66 (2017)

Supervised Learning for Human Action Recognition from Multiple Kinects

Hao Wang[1], Christel Dartigues-Pallez[2(✉)], and Michel Riveill[2]

[1] East China Normal University, Laboratoire d'Informatique Signaux et Systèmes de Sophia Antipolis (i3S), Antipolis, France
wanghao@stu.ecnu.edu.cn
[2] Laboratoire d'Informatique Signaux et Systèmes de Sophia Antipolis (i3S), Antipolis, France
christel.dartigues-pallez@unice.fr, michel.riveill@univ-cotedazur.fr

Abstract. The research of Human Action Recognition (HAR) has made a lot of progress in recent years, and the research based on RGB images is the most extensive. However, there are two main shortcomings: the recognition accuracy is insufficient, and the time consumption of the algorithm is too large. In order to improve these issues our project attempts to optimize the algorithm based on the random forest algorithm by extracting the features of the human body 3D, trying to obtain more accurate human behavior recognition results, and can calculate the prediction results at a lower time cost. In this study, we used the 3D spatial coordinate data of multiple Kinect sensors to overcome these problems and make full use of each data feature. Then, we use the data obtained from multiple Kinects to get more accurate recognition results through post processing.

1 Introduction

Human Action Recognition (HAR) is an active research topic in Computer Vision and a very popular and useful task in various fields. It especially plays an important role in people's daily life. Human fall detection systems are very often needed for many people in today's aging population including the elderly and people with special needs such as the disabled, as fall is the main cause of injury-related death for elderly people [1,2]. Automatic detection of human fall is then a key issue in health management systems. At the same time, HAR is also used in smart home, security video surveillance security and Tele-immersion System. Different approaches are used to build human fall detection systems, including wearable based devices, non-wearable sensors, and vision-based system. Wearable based devices such as accelerometers and gyroscopes are highly preferred by engineers and doctors [3–5]. However, methods based on those equipment have some shortcomings due to the lack of understanding of context and the ability to extract information features [6]. Wearable devices often generate too many false alarms, and wearable devices can also cause inconvenience to people's lives, resulting in a reduced willingness for elderly equipment. There are

Y. Nah et al. (Eds.): DASFAA 2020 Workshops, LNCS 12115, pp. 32–46, 2020.
https://doi.org/10.1007/978-3-030-59413-8_3

also sensor devices that do not need to be worn, installed in a room environment such as floor vibration sensors. These non-wearable devices eliminate the trouble of wearing, but it is still difficult to satisfy people on accuracy. Therefore, the scheme based on visual devices such as cameras has become an applicable choice because it can acquire more human motion information and has a wide range of detection. For these reasons, vision-based devices have higher accuracy in the daily behavior classification of the human body. In the past, there were many works based on vision-based devices for human action recognition. But due to the influence of variations of people, illumination and viewpoint, activity phase and occlusions, there will still be more false positives [7–10].

The emergence of Microsoft Kinect has opened up new opportunities for solving these problems. The Kinect sensor combines a special infrared light source to capture depth information. Meanwhile, Kinect's SDK can generate human skeleton data. RGB data can provide important features of the human body's appearance, but also has a larger range of acquisition. However, the calculation of RGB data features always requires a lot of time, which is not well adapted to the needs of daily life. To avoid this problem, we mainly use Skeleton data and depth data, as it helps to more accurately identify human actions. Skeleton data is mainly composed of scalar vectors, and the calculation speed is thus very fast. Kinect sensors are also limited by the measurement angle and distance range, and they are also affected by noise: people may exceed the monitoring range when falling, resulting in unsatisfactory action recognition. In order to solve this problem, many studies consider multiple Kinect to capture human action from different angles and distances. We can then integrate the data obtained for the final prediction results. The prediction is done thanks to a learning algorithm. In our research, we developed a successful approach based on Random Forest [11].

We will first discuss about existing HAR works and we will present the learning algorithm on which we based our research (Random Forest). Secondly, we will present our methodology and we will present in a third part the dataset we choose and our experiments.

2   Related Work

2.1   RGB-Based Work

The academic community has rich research on human action recognition. We mentioned solutions mainly for wearable devices, non-wearable environmental sensors, and vision-based devices. We mainly discuss human body recognition based on visual information of RGB cameras or Kinect cameras to obtain accurate contour and depth information of the human body. Among all the existing methods, extracting features using RGB image data is the most popular approach. The RGB camera is inexpensive, and it has also spawned many datasets based on RGB images. Most of the methods based on RGB image detection of human action first need to detect the human body area, draw a border of the human body contour and then extract the behavior characteristics of the human body in the border. The work in [7] proposed to use variations in silhouette area that are obtained from only one camera. They use a simple background separation method to acquire the silhouette and find that the proposed feature is view-invariant. And the work in [8] used Support Vector Machine (SVM) for classification. The foreground human silhouette is extracted via background modeling and tracked throughout the video sequence. The human body is represented with ellipse fitting. Then, the shape deformation quantified from the fitted silhouettes is used as the features to distinguish different postures of the human. Finally, they classify different postures via a multi-class SVM and a context-free grammar-based method. The work in [12] tried to estimate 3D human pose from a sequence of monocular images. This paper presents a Recurrent 3D Pose Sequence Machine(RPSM) to automatically learn the image-dependent structural constraint and sequence-dependent temporal context by using a multi-stage sequential refinement. And get better results on Human3.6m [13] and HumanEva-I dataset [14].

Joao Carreira, Andrew Zisserman used deep Convolutional Networks (ConvNets) in 2014 to identify human action in the video [9]. They attempted to capture the complementary information on appearance from static frames and motion between frames. A dual-stream ConvNet architecture with spatial and temporal networks was proposed. Under the wired training data, ConvNet trained on multi-frame dense optical flow can achieve excellent performance: 88% accuracy was obtained on UCF-101 dataset [10] and 59.4% accuracy on HMDB-51 dataset [15]. They did further work [16] based on this. The original model was upgraded, and the 3D convolutional neural network can be constructed by computing features from both spatial and temporal dimensions. The training was re-trained on the new training set Kinetics Human Action Video dataset. The result achieved 97.9% accuracy on UCF-101 dataset and 80.2% on the HMDB-51 dataset. The use of convolutional neural networks requires high hardware (GPU) and extended training time. It needs to adjust parameters to get the best model. The prediction accuracy may not be guaranteed after replacing new dataset.

Some other works based on multiple RGB cameras are also very instructive. The work in [17] used two cameras to acquire graphics. They raised a multi-view fall detection system where a layered Hidden Markov Model models motion. In [18], they compared two approaches for the detection of falls based on multiple cameras: the early fusion approach and the late fusion approach. In the early fusion approach, multiple camera views are combined to reconstruct the 3D voxel volume of the human. Based on semantic driven features, fall detection is done on this 3D volume, whereas in the late fusion fall detection is done in 2D, and each camera decides on its own. If a fall occurred, the system will combine these individual decisions into an overall decision. Early approaches provided a better performance, while late fusion required less computing power and was easier to handle. All those approaches are very interesting and challenging, but they are often time consuming and very dependant of the lighting conditions. Other HAR researches voluntarily adopt simpler approaches based on the skeleton.

2.2 Skeleton-Based Work

With the advent of the Kinect camera, the efficient RGB-D sensor provides a new direction for human action recognition. In addition to RGB graphics, the Kinect camera provides depth and skeleton information independent of lighting conditions. In [18], the author used the depth pattern to extract human body image boundaries. Then they calculated the curvature dimension spatial characteristics of the human contour and applied the extreme learning machine to classify the different actions. The work in [19] used the hierarchical recurrent neural network (RNN) to perform motion recognition on 3D skeleton data. They divided the human skeleton into five parts according to the human physical structure and then separately feed them to five RNN subnets. They get excellent performance, but this method encounters overfitting problems. In [20], the authors chose a set of key-pose-motifs for each action class. They classified a sequence by matching it to the motifs of each class and selecting the class that maximizes the matching score. The work in [21] used an angular representation of the skeleton joints to describe each pose. They used those descriptors to identify key poses through a multi-class SVM. The gesture is then labeled from the key pose sequence with a decision forest.

For the fall detection problem, there are more specific options to choose. Two fall detection algorithms are proposed in [22]. One determines whether a drop has occurred by a single frame. The second uses time-series data to distinguish falls and slowly lay down on the floor. In [23], they tried to explore secondary features (angle and distance), focusing on the correlation between joints and the boundary of this correlation. The authors mainly focused on the angle of the joints on the legs, and the distance from the floor to several important joint points. The algorithm is simple, but the prediction results are unstable due to the quality of joint tracking. Author of [24] considered that Kinect could not track all joints correctly. They defined and computed three features (distance, angle, velocity) on only several vital joints. Then they used SVM to analyzed ten specific actions with good results.

Trying to combine RGB data with skeleton data is also an effective method. There is a novel method in [25] which uses skeleton data to obtain the 3D bounding box of the human body. It then measures the velocity based on the contraction or expansion of the width, height, and depth of the 3D bounding box. The authors of [26] creatively installed the camera on the ceiling. The human head-to-ceiling distance is an important feature that combines the application of the accelerometer with the K-Nearest-Neighbour (KNN) classifier for identification. The work in [27] used a tri-axial accelerometer to indicate the potential fall as well as to indicate whether the person is in motion. If the measured acceleration is higher than an assumed threshold value, the algorithm extracts the skeleton, calculates the features and then executes the SVM-based Classifier to authenticate the fall alarm. A similar work in [28] is also using KNN, where an accelerometer is used to indicate a potential fall, and the Kinect sensor is used to authenticate an eventual fall alert. Only the depth image captured during the possible fall is processed.

It is also quite skillful to know how to combine the information obtained by multiple cameras. Earlier fusion and late fusion are mentioned. The work in [29] tried to use a new cross-view action representation. They propose a method effectively express the geometry, appearance, and motion variations across multiple viewpoints with a hierarchical compositional model. They used 3D skeleton data acquired from Kinect to train, Northwestern-UCLA Multiview Action3D Dataset and dataset MSR-DailyActivity3D Dataset [30]. Then they tested the model with unknown 2D video. They succeeded to use the cross-view to improve the accuracy and robustness of action recognition.

We mainly focus on the work in [31] using the combination of RGB and skeleton data. They installed seven Kinects with different angles in the room. They used the skeleton data to obtain the vertical velocity and the height of the human body from the ground. If the Kinect tracking fails and does not generate enough skeleton data, the features are extracted from the continuous RGB data. The human action recognition is performed based on the SVM classifier. Finally, the results of the seven Kinect data processing are combined. This method also achieved an accuracy of 91.5%, but we can observe two main problems: processing RGB images still require a lot of calculations and skeleton data are not used enough. In many cases, the use of RGB images tends to cause users' concerns about privacy issues. All those studies show that using skeleton instead of purely RGB data is a good solution. Combining the skeleton with another classifier such as the Random Forest also shows interesting results [11] on the MSR-DailyActivity3D Dataset. In this study the vector representing a moment in the flow of the data is composed of all the coordinate of the joints of the skeleton and all the distances and angles between the joints. Several consecutive moments are combined in one vector in order to fully describe an action. Thanks to all those works and especially the last one, we choose in this study to consider Random Forest as main classifier for our work.

3   Technical Overview

In this section, we describe the basic concepts and characteristics of Microsoft Kinect camera and explain the skeleton data generated by this equipment. We then explain the principle of the Random Forests and explai why they are suitable for our human behavior recognition task.

3.1   Microsoft Kinect

Kinect is a motion-sensing input device by Microsoft for the Xbox360 video game console and Windows PCs. Kinect with full skeleton mode can track a person's actions and generate 20 joints as the skeleton data [32]. Each joint include the value of (x, y, z) in 3-dimensional space. Figure 1 shows the 20 joints of the human body. Our work uses the data generated by full skeleton mode.

The skeleton data obtained from Kinect is 20 key joints of the body. Each joint is a 3- dimensional vector. The data volume of 20 joints is not enough to

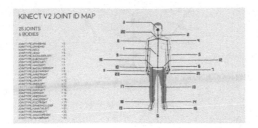

Fig. 1. 20 joints generated by Kinect

support the requirements of training data for machine learning algorithms. We refer to the method in [27] to calculate angles, distances, and other information through different joints. In this way, more human body motion features can be extracted, and the amount of data is greatly enriched, which is very helpful for the classification algorithm.

3.2 Random Forest

Random Forest have been first formally introduced in [33]. In this paper, Leo Breiman define a Random Forest as a multi-classifier composed of a set a decision trees. The method defined by Breiman is called Forest-RI (Random Forest - Random Input) and is still a very popular approach.

In our work, we use the R-package Random Forest v4.6-14 to implement the Random Forest. This package implements Breiman's Random Forest algorithm (based on Breiman and Cutler's original Fortran code) for classification and regression.

4 Methodology

In this section, we present our approach of classification of HAR from a multi-Kinect dataset with Random Forest. We first describe the vector created from the raw dataset. We then describe our innovative approach based on two important points: the cutting of the skeleton into five significant subparts and the development of a hierarchical Random Forest algorithm. We will end this part by describing how we managed the multi-views data.

4.1 Feature Vector

Since we only use 3D skeleton data, we first consider 20 joints (characterized by (x, y, z) coordinates) in a three-dimensional vector. This vector doesn't contain enough data to fully classify different actions. Similarly to the work done in [11], we have augmented our feature vector by compute and add all possible distances/angles between all possible pair/triplet of joints. This process ended

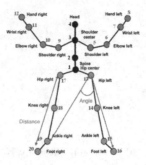

Fig. 2. Distance and Angle of skeleton joints

to a feature vector of 3610 values: 3420 angle values followed by 190 distance values.

As shown in Fig. 2, a distance data can be generated between any two joints, and three angle vectors can be calculated between any three joints. This feature vector will be represented as FV in this paper. Each frame can generate one-row FV like Fig. 3.

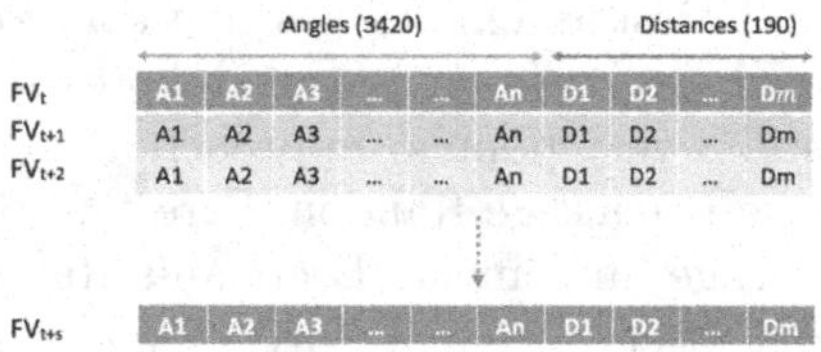

	Angles (3420)						Distances (190)			
FV$_t$	A1	A2	A3	...	...	An	D1	D2	...	Dm
FV$_{t+1}$	A1	A2	A3	...	...	An	D1	D2	...	Dm
FV$_{t+2}$	A1	A2	A3	...	...	An	D1	D2	...	Dm
FV$_{t+s}$	A1	A2	A3	...	...	An	D1	D2	...	Dm

Fig. 3. Feature vector of angles and distances

At the same time, the action of the human body is a dynamic process; we need to acquire the temporal features of an action at the same time. For each frame with time index t, we extract the pairwise relative position features by taking the difference between the position of joint at time t and that of others frame with time index t+1:

$$Diff2_t = FV_{t+1} - FV_t \tag{1}$$

This new feature $Diff2_t$ provides some temporal information by varying the angle and distance between two consecutive frames. But calculating the difference between two adjacent frames is not robust. We need to rich temporal information. In the work of [34, 35], there is a Spatial Pyramid approach. We set a 10-frame sliding window, and calculate the difference between different number of frames, 2,5 and 10 frames. Then we get the $Diff5_t$ $Diff5M_t$ $Diff10_t$ represent the action movements in 5 and 10 frames.

$$Diff5_t = FV_{t+4} - \sum_{i=t}^{i=t+3} FV_i \qquad (2)$$

$$Diff5M_t = FV_{t+8} - \sum_{i=t+4}^{i=t+7} FV_i \qquad (3)$$

$$Diff10_t = FV_{t+9} - \sum_{i=t}^{i=t+8} FV_i \qquad (4)$$

In a 10-frame sliding window, the differences between each pair of consecutive poses will be sum up into two Diff5 feature vectors, one from t to t+4 and another one from t+5 to t+9. In order to preserve the coverage we had with the Diff2 feature vector and the overlapping sliding window. We also define a "Middle" Diff5 feature vector by calculating the middle 5 position and averaging the Diff2 at these positions [11]. The Fig. 4 shows the consistency of each Diff feature vector.

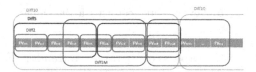

Fig. 4. Sliding window and definition of Diff2, Diff5, Diff5M, Diff10 feature vector

Using 10 frames we can obtain 9 Diff2, 2Diff5, 2Diff5M and 1 Diff10 features. At last, we can get 46930 values for every 10 frames. We will use this FV-46930 as the input of the Random Forest. The acquisition rate of CMDFALL is 20 Hz. Therefore, the Kinect camera will capture 20 frames in one second. We set 10 frames as the sliding window, which is reasonable because most fall action in the dataset are completed in 0.5 s.

4.2 Decompose Whole Body Data into Subparts

Training a Random Forest with whole-body data points does not necessarily yield good results because some human actions do not lead to whole-body movements. Moreover, this leads to very long computation time. In order to reduce preprocessing time and learning time we decompose the human body into five distinct subparts: left arm, right arm, left leg, right leg and the upper of the body. On our server machine (2 Processors Intel Xeon X5675 at 3,06 GHz and 24 GB RAM), calculating the feature vectors (distances and angles) by whole body (20 joints) from more than 400 files will take more than 24 h. In subpart mode, it only consumes few minutes.

For each subpart of the body, a Random Forest is build. Each subpart contains 4 joints, and we also calculate the distances and angles. The 4 joints could generate a feature vector with 6 distance values and 12 angle values. For each subpart-Random Forest, we calculate a prediction score and normalize the five obtained scores to get a percentage of the prediction. Further, we connect the prediction percentage generated by each subpart-Random Forest with the distances/angles vector to form a new feature vector (FV-190) with 190 values. Then we use the FV-190 to build a new Random Forest. The new feature vector is shown in Fig. 5.

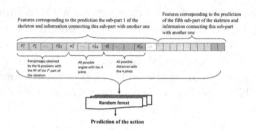

Fig. 5. New feature vector with 190 values

Figure 6 shows the entire process of Hierarchical Random Forests. At first, we decompose the human body into five subparts. Secondly we use partial skeleton data to build five subpart-Random Forests and obtain five prediction scores. We then compose the new feature vector by concatenating the prediction scores obtained in the preceding step and the distances/angles feature of each subpart. A new Random Forest is finally built thanks to this new vector to obtain the final decision result.

4.3   Multi-views Skeleton Data

For this study we restricted ourselves to 3 Kinect cameras installed from 3 different angles: skeleton data of the same action are recorded by 3 Kinect cameras at the same time. We can easily generalized to more than 3 Kinects. Integrating the 3 views skeleton data can help us to improve the precision of action recognition. We provide three strategies and validated the predictions of these different strategies through experiments.

Strategy 1: Merge 3 Kinect vectors into 1 vector and build 1 RF

In our first strategy, we consider the feature vectors generated by each Kinect and we merge those 3 vectors into one. We then get a bigger feature vector with 140790 values. In this way each input sample is characterized by more features. We use all the obtained vectors for our dataset and use them to build a single Random Forest.

Strategy 2: Merge 3 Kinect vectors in 3 vectors and build 1 RF In this strategy, also called early fusion strategy, we start with the three feature vectors

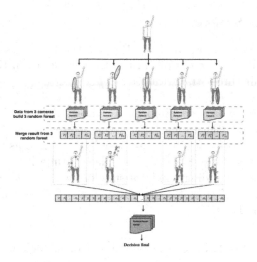

Fig. 6. Process of hierarchical random forest

generated each Kinect and we use all of them to build only one dataset. This dataset is then used to build a single Random Forest. In this way, multi-views only able to add the training samples.

Strategy 3: Consider 3 Kinect vectors and build 3 RFs The last strategy, also called late fusion strategy, consists in building a dataset for each Kinect specific vector. Each dataset is then used to build a Random Forest and the decisions of each RF are then combined.

5 Experiment

We test different steps of our algorithm on the CMDFALL dataset and we compared our result to state-of-art methods in [31].

5.1 Dataset and Setup

We use CMDFALL as the data set for our experimentation to test our approach. Calculating the distance and angle of the skeleton and calculating the difference between consecutive frames requires a lot of calculations. We thus perform operations on a server. The hardware configuration is: Dell PowerEdge R710 Rack, 2 Processors Intel Xeon X5675 at 3,06 GHz, 6 Cores, 12MB cache memory, 24 GB RAM DDR3-1333 MHz, 2 hard disks Hot Plug 600GB SAS 6Gbit/s 15000tr/min with RAID 0 for performance. At our data scale, building the Random Forest does not need high hardware requirements. Building a Random Forest on the subparts of the skeleton takes about a few minutes, and it takes about one to two hours to build a Random Forest of the whole body skeletons.

5.2 Whole Body Skeleton

We test the whole-body mode at first. We use the 20 joints skeleton data to build the model and fuse the data from the three cameras in the different ways corresponding to our three strategies.

Strategy 1: Merge 3 Kinect vectors in 3 vectors and build 1 RF In the first method, we merged the data acquired by the three Kinects into a single vector. Each Kinect's data for each frame generates a vector of 46,930 values that are combined into a vector of 140,790 values. 5 and 10 are used as parameters of the sliding window, respectively. Since the calculation time is too long, we do not calculate the number of sliding windows smaller than 5. At the same time, we used different number of trees for the forest: 500, 1000 and 1500. The specific results are shown in Table 1.

Table 1. Class error rate of 3 views in one row

Tree number	500	1000	1500	2000
Sliding window 10	37.43%	36.05%	33.84%	39.49%
Sliding window 5	29.43%	28.14%	27.98%	27.30%

Result in Table 2 shows there are not obvious improvement for class classification accuracy. But precision of detection the fall action improves a lot. Only for fall action class, the precision is about 90

Table 2. Merged class error rate of 3 views in one row

Tree number	500	1000	1500	2000
Sliding window 10	35.23%	32.65%	31.06%	35.88%
Sliding window 5	29.58%	27.65%	27.78%	25.58%

Strategy 2: Merge 3 Kinect vectors in 3 vectors and build 1 RF In the second method, we treat all the views side by side. Three batch of skeleton data from different Kinects will be used as input samples together. Similarly, we get the following results in Table 3. There is no obvious improvement in the results, even worse. According to the analysis of the classification results, the same actions from different views are not easily classified into same class. Explain that the feature values of the actions captured by different views obtained by our method have large differences. We believe that there should be better means to obtain more relevant feature from different views.

Strategy 3: Consider 3 Kinect vectors and build 3 RFs In the third method, we build three independent Random Forests using data from different Kinects. After the test data input, we add the predicted scores of the three Random

Table 3. Class error rate of 3 views in 3 rows

Tree number	500	1000	1500	2000
Sliding window 10	40.52%	39.10%	39.83%	38.78%
Sliding window 5	38.26%	36.99%	37.54%	25.58%

Forest outputs, and the highest score is the final result. The final classification results are shown in the Table 4. The superposition of their respective prediction results from 3 independent Random Forests, did not significantly improve the precision rate. According to the detailed classification results, in some cases, the 3 Random Forests will be wrong in predicting the same class.

Table 4. Class error rate of 3 views defining 3 random forests

Tree number	500	1000	1500	2000
Sliding window 10	34.13%	32.28%	37.64%	36.43%
Sliding window 5	28.23%	27.64%	26.12%	28.04%

5.3 Subpart Body and Hierarchical Random Forest

Using whole body skeleton data does not get ideal result. In order to achieve higher accuracy, we try to break down the human body into 5 parts, to build the five sub-Random Forests separately and to integrate the prediction results and combine them to build a new Random Forest (Hierarchical Random Forest). We tested each with a sliding window of 10. Number of trees in random forest is 500. We randomly choose X% samples as training set and $(1-X)$ % as testing set. Using training set to define the Hierarchical Random Forest and input the testing set to the global Random Forest to get the prediction result. The results are much better than for the whole body. The result is shown in Table 5. We set X as 80, 60 and 40. Kinect3, Kinect4 and Kinect5 means the data is captured by the Kinect cameras 3, 4 and 5.

Table 5. Classification precision of Hierarchical random forest

Training percent	80%	60%	40%
Kinect 3	99.10%	97.08%	95.40%
Kinect 4	99.55%	97.60%	97.40%
Kinect 5	99.10%	96.81%	95.82%

When using 80% dataset as training set, we could get precision over 99% in individual view. And our method is simpler and faster in computing, using only skeleton data. It consumes about 5 min to calculating feature vector (distances and angles) from 418 files, each files. Building random forest will take less than 5 min.

6   Conclusion

This paper proposes a new human action detection method that uses only three-dimensional skeleton data. Without the using RGB images or motion velocity information collected by other wearable sensors such as accelerometer and gyroscopes. Using only skeleton data can reduce the amount of computation, while also avoiding the troubles of wearing devices. We constructed a Hierarchical Random Forest with five subparts of the whole body skeleton decomposition. Subpart mode effectively reduce the time consumption of traditional algorithms and Hierarchical Random Forest greatly improve the precision of classification. We get the most static features by calculating all possible angles and distances between each joint. The temporal characteristics are extracted by calculating the difference directly between adjacent frames. We tested our approach on the CMDFALL data set and the results are satisfactory for the whole body but they are better with the subparts combined with the Hierarchical Random Forest, with an average classification precision 98.5% on CMDFALL.

Since in the CMDFALL dataset, fall actions only happend in a part of a group of action frames. We can try to detect the large movement of the human body through the accelerometer and get the time information of the action to extract data for the corresponding time period. This way, the accurate skeleton data corresponding to the fall process can be obtained and results of action recognition may be more accurate.

Acknowledgements. This work was supported by NSFC grants (No. 61532021 and 61972155).

References

1. Griffiths, C., Rooney, C., Brock, A.: Leading causes of death in England and wales-how should we group causes. Health Stat. Q. **28**(9), 6–7 (2005)
2. Nizam, Y., Haji Mohd, M.N., Abdul Jamil, M.M.: Classification of human fall from activities of daily life using joint measurements. J. Telecommun. Electron. Comput. Eng. (JTEC) **8**(4), 145–149 (2016)
3. Bourke, A.K., O'brien, J.V., Lyons, G.M.: Evaluation of a threshold-based tri-axial accelerometer fall detection algorithm. Gait Posture **26**(2), 194–199 (2007)
4. Bagalà, F., et al.: Evaluation of accelerometer-based fall detection algorithms on real-world falls. PLoS ONE **7**(5), e37062 (2012)
5. Yang, C.-C., Hsu, Y.-L.: A review of accelerometry-based wearable motion detectors for physical activity monitoring. Sensors **10**(8), 7772–7788 (2010)
6. Gioanni, L., Dartigues-Pallez, C., Lavirotte, S., Tigli, J.-Y.: Using random forest for opportunistic human activity recognition: a complete study on opportunity dataset. In: 11èmes journées francophones Mobilité et Ubiquité, Ubimob 2016, Lorient, France, 5 July 2016 (2016)

7. Mirmahboub, B., Samavi, S., Karimi, N., Shirani, S.: Automatic monocular system for human fall detection based on variations in silhouette area. IEEE Trans. Biomed. Eng. **60**(2), 427–436 (2013)

8. Feng, W., Liu, R., Zhu, M.: Fall detection for elderly person care in a vision-based home surveillance environment using a monocular camera. SIVIP **8**(6), 1129–1138 (2014)

9. Simonyan, K., Zisserman, A.: Two-stream convolutional networks for action recognition in videos. In: Advances in Neural Information Processing Systems 27: Annual Conference on Neural Information Processing Systems 2014, Montreal, Quebec, Canada, 8–13 December 2014, pp. 568–576 (2014)

10. Soomro, K., Zamir, A.R., Shah, M.: UCF101: a dataset of 101 human actions classes from videos in the wild. CoRR, abs/1212.0402 (2012)

11. Aly Halim, A., Dartigues-Pallez, C., Precioso, F., Riveill, M., Benslimane, A., Ghoneim, S.A.: Human action recognition based on 3d skeleton part-based pose estimation and temporal multi-resolution analysis. In: 2016 IEEE International Conference on Image Processing, ICIP 2016, Phoenix, AZ, USA, 25–28 September 2016, pp. 3041–3045 (2016)

12. Lin, M., Lin, L., Liang, X., Wang, K., Cheng, H.: Recurrent 3D pose sequence machines. In: 2017 IEEE Conference on Computer Vision and Pattern Recognition, CVPR 2017, Honolulu, HI, USA, 21–26 July 2017, pp. 5543–5552 (2017)

13. Ionescu, C., Papava, D., Olaru, V., Sminchisescu, C.: Human3.6m: large scale datasets and predictive methods for 3D human sensing in natural environments. IEEE Trans. Pattern Anal. Mach. Intell. **36**(7), 1325–1339 (2014)

14. Sigal, L., Black, M.J., HumanEva: synchronized video and motion capture dataset for evaluation of articulated human motion., Technical report (2006)

15. Kuehne, H., Jhuang, H., Garrote, E., Poggio, T.A., Serre, T.: HMDB: a large video database for human motion recognition. In: IEEE International Conference on Computer Vision, ICCV 2011, Barcelona, Spain, 6–13 November 2011, pp. 2556–2563 (2011)

16. Carreira, J., Zisserman, A.: Quo Vadis, action recognition? A new model and the kinetics dataset. In: 2017 IEEE Conference on Computer Vision and Pattern Recognition, CVPR 2017, Honolulu, HI, USA, 21–26 July 2017, pp. 4724–4733 (2017)

17. Thome, N., Miguet, S., Ambellouis, S.: A real-time, multiview fall detection system: a LHMM-based approach. IEEE Trans. Circ. Syst. Video Technol. **18**(11), 1522–1532 (2008)

18. Ma, X., Wang, H., Xue, B., Zhou, M., Ji, B., Li, Y.: Depth-based human fall detection via shape features and improved extreme learning machine. IEEE J. Biomed. Health Inform. **18**(6), 1915–1922 (2014)

19. Du, Y., Wang, W., Wang, L.: Hierarchical recurrent neural network for skeleton based action recognition. In: IEEE Conference on Computer Vision and Pattern Recognition, CVPR 2015, Boston, MA, USA, 7–12 June 2015, pp. 1110–1118 (2015)

20. Wang, C., Wang, Y., Yuille, A.L.: Mining 3D key-pose-motifs for action recognition. In: 2016 IEEE Conference on Computer Vision and Pattern Recognition, CVPR 2016, Las Vegas, NV, USA, 27–30 June 2016, pp. 2639–2647 (2016)

21. Miranda, L., Vieira, T., Morera, D.M., Lewiner, T., Vieira, A.W., Campos, M.F.M.: Online gesture recognition from pose kernel learning and decision forests. Pattern Recogn. Lett. **39**, 65–73 (2014)

22. Kawatsu, C., Li, J., Chung, C.J.: Development of a fall detection system with microsoft kinect. In: Kim, J.H., Matson, E., Myung, H., Xu, P. (eds.) Robot Intelligence Technology and Applications 2012 Advances in Intelligent Systems and Computing, vol. 208, pp. 623–630. Springer, Heidelberg (2012). https://doi.org/10.1007/978-3-642-37374-9_5910.1007/978-3-642-37374-9_59

23. Flores-Barranco, M.M., Ibarra-Mazano, M.-A., Cheng, I.: Accidental fall detection based on skeleton joint correlation and activity boundary. In: Bebis, G., et al. (eds.) ISVC 2015. LNCS, vol. 9475, pp. 489–498. Springer, Cham (2015). https://doi.org/10.1007/978-3-319-27863-6_45

24. Tran, T.-T.-H., Le, T.-L., Morel, J.: An analysis on human fall detection using skeleton from microsoft kinect. In: 2014 IEEE Fifth International Conference on Communications and Electronics (ICCE), pp. 484–489, July 2014

25. Mastorakis, G., Makris, D.: Fall detection system using kinect's infrared sensor. J. Real-Time Image Proc. **9**(4), 635–646 (2014)

26. Kepski, M., Kwolek, B.: Fall detection using ceiling-mounted 3D depth camera. In: VISAPP 2014 - Proceedings of the 9th International Conference on Computer Vision Theory and Applications, vol. 2, Lisbon, Portugal, 5–8 January 2014, pp. 640–647 (2014)

27. Kwolek, B., Kepski, M.: Human fall detection on embedded platform using depth maps and wireless accelerometer. Comput. Methods Programs Biomed. **117**(3), 489–501 (2014)

28. Kwolek, B., Kepski, M.: Improving fall detection by the use of depth sensor and accelerometer. Neurocomputing **168**, 637–645 (2015)

29. Wang, J., Nie, X., Xia, Y., Wu, Y., Zhu, S-C.: Cross-view action modeling, learning, and recognition. In: 2014 IEEE Conference on Computer Vision and Pattern Recognition, CVPR 2014, Columbus, OH, USA, 23–28 June 2014, pp. 2649–2656 (2014)

30. Wang, J., Liu, Z., Wu, Y., Yuan, J.: Mining actionlet ensemble for action recognition with depth cameras. In: 2012 IEEE Conference on Computer Vision and Pattern Recognition, Providence, RI, USA, 16–21 June 2012, pp. 1290–1297 (2012)

31. Tran, T.-H., Le, T.-L., Hoang, V.-N., Hai, V.: Continuous detection of human fall using multimodal features from kinect sensors in scalable environment. Comput. Methods Programs Biomed. **146**, 151–165 (2017)

32. Shotton, J., et al.: Real-time human pose recognition in parts from single depth images. In: Proceedings of the 2011 IEEE Conference on Computer Vision and Pattern Recognition, CVPR 2011, pp. 1297–1304. IEEE Computer Society, Washington (2011)

33. Breiman, L.: Random forests. Mach. Learn. **45**(1), 5–32 (2001)

34. Wang, J., Liu, Z., Wu, Y., Yuan, J.: Mining actionlet ensemble for action recognition with depth cameras. In: 2012 IEEE Conference on Computer Vision and Pattern Recognition, pp. 1290–1297. IEEE (2012)

35. Lazebnik, S., Schmid, C., Ponce, J.: Beyond bags of features: spatial pyramid matching for recognizing natural scene categories. In: 2006 IEEE Computer Society Conference on Computer Vision and Pattern Recognition (CVPR 2006), vol. 2, pp. 2169–2178. IEEE (2006)

Discovery of Chasing Patterns in Trajectory Data

Huaqing Wu[1], Weizhuo He[1,2], Qizhi Liu[1(✉)], and Yu Yang[3]

[1] State Key Laboratory for Novel Software Technology, Nanjing University, Nanjing, China
lqz@nju.edu.cn
[2] HUAWEI Corporation, Nanjing, China
[3] Pancar Technology Ltd., Nanjing, China

Abstract. Trajectory data contains abundant temporal and spatial information of moving objects. By calculating and analyzing trajectory data, behavior patterns of mobile objects can be found. Among these behavior patterns the discovery of chasing patterns is interesting which will provide effective services for intelligent urban traffic management, criminal investigation, environmental monitoring and other application fields. However, existing chasing pattern discovery methods only compare the local similarity of trajectories, and do not consider the potential significance of stay points or change in speed. This article proposes a new algorithm for chasing pattern discovery based on the stay point detection technology. An optimized stay point detection algorithm is also designed to improve the distance calculation method. The algorithms were evaluated with real and simulation trajectory data and achieved very good results which have higher accuracy and recall rate compared with the existing chasing pattern discovery algorithms.

Keywords: Trajectory data · Chasing patterns · Stay point detection · Great circle distance

1 Introduction

The widespread use of modern positioning technology has produced a large amount of trajectory data. The trajectory data contains rich spatio-temporal information. An in-depth analysis of the spatio-temporal information in the trajectory data can discover the behavior patterns of moving objects, and can produce interesting and beneficial application effects in multiple fields. For example, Alvares et al. proposed an algorithm to detect avoidance behavior pattern that identifies when a moving object is avoiding specific spatial regions, such as security cameras [1]. Jeung et al.'s research on animal group movement pattern discovery can be used for migration event recognition [2]. Siqueira and Bogorny presented formal definitions and an algorithm to find chasing behavior in moving object trajectories [3]. Kang et al. used the chasing pattern to explain the behavioral characteristics of a class of players in the football game, and formulated a new strategy for subsequent games by observing the chasing pattern of the athletes [4]. Among various behavior patterns, we believe that chasing pattern is a very important

© Springer Nature Switzerland AG 2020
Y. Nah et al. (Eds.): DASFAA 2020 Workshops, LNCS 12115, pp. 47–59, 2020.
https://doi.org/10.1007/978-3-030-59413-8_4

pattern, and its discovery technology can be used in applications such as urban traffic management, driverless, security, etc., and can prevent some pre-planned abduction or terrorism to reduce the occurrence of potentially malignant events.

The chasing pattern involves many spatio-temporal information such as the position, time stamp, speed, direction and change trend of the pair of moving objects in the trajectory data, and its discovery process is relatively complicated. This article specializes in chasing pattern discovery methods, and it is the first method that based on stay point detection technology. Because chasing and chased target often share similar staying characteristics, converting the original GPS trajectory of the moving object into a trajectory with stay points will not only help the discovery of the chasing pattern, improving the recall rate, but also help to find more complete chasing pattern. Meanwhile, it can also be used to compress the trajectory data, reducing the calculation amount of pattern discovery. The main contributions of this article are as follows:

- We establish a chasing pattern discovery model based on the detection of stay points. The model can effectively find chasing patterns in trajectories with large angle changes. The detection of stay points in the model helps to screen out trajectories that are more likely to have chasing patterns. The candidate trajectory set can also improve the efficiency of subsequent calculations.
- We design a chasing pattern discovery algorithm. And the stay point detection algorithm is optimized to improve the accuracy of similar trajectory matching and chasing pattern discovery.
- We verified the effectiveness of the above model and the performance of the algorithms through a series of experiments.

In the rest of this article, we give the definition of the problem and the chasing pattern discovery model in the second section. In the third section, we describe the optimization algorithm for stay point detection and the chasing pattern discovery algorithm based on stay point detection. The fourth section introduces the experimental settings and the results of each experiment. The fifth section introduces related work. Finally, we summarizes the full text and discusses future work directions.

2   Chasing Pattern Discovery Model

A trajectory is the path of a moving object and can be described as a function of time. It corresponds to a set of trajectory point data sequences including time stamps and spatial information [5], where the spatial information generally includes the latitude and longitude of the trajectory points, and sometimes includes the altitude of the point. The sequence often contains other information such as the moving speed when the moving object is passing the trajectory point. In this section we give the definition of the original trajectory used in this article, which helps us design a chasing pattern discovery model.

Definition 1 (Original trajectory). *An original trajectory T is a sequence of trajectory points, that is, $T : p_1 \rightarrow p_2 \rightarrow \ldots \rightarrow p_n$. Each trajectory is uniquely identified by the number tid. The trajectory point $p_i(1 \leq i \leq n)$ is a multivariate group including at least four kinds of information: latitude, longitude, velocity, and timestamp, that is,*

p_i : $\{lat_i, lng_i, speed_i, t_i\}$. *The time interval between two consecutive trajectory points does not exceed the time threshold* Δt, *that is,* $0 < t_{i+1} - t_i < \Delta t$.

2.1 Baseline Model

Given a set of candidate original trajectories $\mathbb{T}c$ and an original trajectory T ($T \notin \mathbb{T}c$), the problem of finding the chasing pattern is to match T' (($T' \in \mathbb{T}c$)) with T one by one. The solution model is:

Given two trajectories T and T', $T : p_1 \to p_2 \to \ldots \to p_n$, $T' : q_1 \to q_2 \to \ldots \to q_m$, with the average velocity v_1, v_2 respectively. When the following conditions are met, we have a chasing where T is chasing T'. That is, there is a chasing pattern between T and T':

(1) For the start and end positions in the two trajectories, there are $distance(p_1, q_1) \le \varepsilon$ and $distance(p_n, q_m) \le \varepsilon$, where ε is a small value;
(2) For the average distance threshold Δd, there exists $\sum_{i=1}^{n} distance(p_i, T_2)/n \le \Delta d$ and $\sum_{j=1}^{m} distance(q_j, T_1)/m \le \Delta d$;
(3) For the time difference tolerance Δtd, there exist $0 \le t_{q_1} - t_{p_1} \le \Delta td$ and $0 \le t_{q_m} - t_{p_n} \le \Delta td$;
(4) For the duration parameter τ, there exists $t_{q_m} - t_{p_1} \ge \tau$;
(5) For the average velocity threshold $\alpha \in [0, 1)$, there exists $(1 + \alpha) \ge (v_1/v_2) \ge (1 - \alpha)$.

Among them, constraint condition (5) is optional. For chasing patterns considering speed, the similarity of the average velocity needs to be calculated.

2.2 Optimization Model

Moving objects usually stop moving when they encounter obstacles, events or places of interest during the journey. Those stay information will be implicit in the trajectory data. For example, a significant reduction in speed can cause several spatial attributes of consecutive trajectory points to be similar, but they still differ in time attributes. By analyzing the trajectory data, several trajectory points corresponding to the stay in the original trajectory can be replaced by one stay point (as shown in Fig. 1).

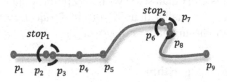

Fig. 1. Stay points

Definition 2 (Stay point). *A stay point $stop_k : p_{k_1} \rightarrow p_{k_2} \rightarrow \ldots \rightarrow p_{k_n}(1 \leq k_1 < k_n \leq n)$ corresponds to a segment of the original trajectory T, where the distance between p_{k_1} and p_{k_n} is less than the threshold Δds, and $t_{k_n} - t_{k_1}$ is greater than the time threshold Δts, corresponding to a geographical area where the moving object stays for a period of time.*

We assume that similar stay behaviors are often found in trajectory pairs with chasing patterns and with the stay point detection technology [6] we can find chasing patterns more effectively in trajectory data. Based on this idea, we design a chasing pattern discovery optimization model, as shown in Fig. 2.

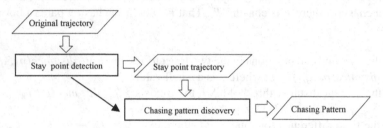

Fig. 2. Frame of the chasing pattern discovery optimization model

In the model, the trajectory segments of all the stay regions in the original trajectory are converted into stay points. Thus forms a new trajectory sequence with stay points, which looks like $T^s : p_1 \rightarrow stop_1 \rightarrow p_4 \rightarrow p_5 \rightarrow stop_2 \rightarrow p_9 \rightarrow \ldots \rightarrow p_n$. When searching for a chasing pattern, we first consider the stay point as an ordinary trajectory point to judge whether there is chasing pattern according to the basic model; then we take the stay points as special trajectory points and use the point-to-point comparison method to identify the true chasing pattern based on the number and distance of the stay points.

3   Chasing Pattern Discovery Algorithm

In the chasing pattern, the track of the chaser is roughly the same as that of the chased, so the chaser may encounter the same events and has similar staying behavior as the chase. Therefore, the stay point is an important feature that can reflect the chasing behavior pattern. Moreover, the trajectory data scale with stay points obtained after using the stay point detection technology is smaller than the original trajectory, which can reduce the amount of calculation for chasing pattern discovery.

3.1   Stay Point Detection Algorithm

Existing stay point detection methods are usually based on the velocity change information in the trajectory [7], or whether the distance between points is less than the distance threshold [8]. This velocity change or distance threshold is only valid in a specific application scenario and a specific time range. In this article, a new stay point detection algorithm (nSPD) is designed to find the chasing patterns considering the time threshold, distance threshold, and the switching times of detection. The algorithm idea is shown in Fig. 3.

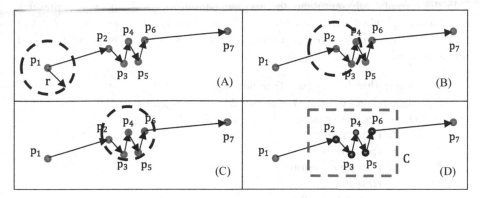

Fig. 3. Stay point detection schematic diagram

Given a trajectory $T : p_1 \rightarrow p_2 \rightarrow \ldots \rightarrow p_7$, we first calculate the physical distance between each point and the point behind it in turn, until the distance between this point and a subsequent point exceeds a given distance threshold r. In Fig. 3(A), $dist(p_1, p_2) > r$ and no other points that constitute a stay point with p_1 are found, so the center point is moved to p_2, as shown in Fig. 3(B). And $dist(p_2, p_3) < r$, $dist(p_2, p_4) < r$, $dist(p_2, p_5) > r$, i.e. $dist(p_2, p_3)$ meets the distance threshold as well as $dist(p_2, p_4)$does, so if the time interval between p_2 and p_4 is not less than the time threshold Δt_s, i.e. $interval(t_2, t_4) \geq \Delta t_s$, we take p_2, p_3, p_4 as a small cluster and move the center to p_4, as shown in Fig. 3(C). Then $dist(p_4, p_5) < r$, $dist(p_4, p_6) < r$, so p_2, p_3, p_4, p_5, p_6 compose a cluster, i.e. the stay point C, as shown in Fig. 3(D). The coordinates of the stay point are obtained by calculating the geometric mean of the coordinates of those points. After that, the original trajectory sequence $\{p_1 \rightarrow p_2 \rightarrow p_3 \rightarrow p_4 \rightarrow p_5 \rightarrow p_6 \rightarrow p_7\}$ becomes the sequence $\{p_1 \rightarrow stop_c \rightarrow p_7\}$. The original 7 points are reduced to 3 points.

In general, for a series of consecutive points $P : \{p_{k_1} \rightarrow p_{k_2} \rightarrow \ldots \rightarrow p_{k_n}\}(1 \leq k_1 < k_n \leq n)$ in an original GPS trajectory $T : p_1 \rightarrow p_2 \rightarrow \ldots \rightarrow p_n$, the nSPD algorithm determines whether P can be regarded as a stay point $stop_k$ according to three parameters, i.e. the time threshold Δt_s, the distance threshold r and the number of times of detection threshold θ_{times}. When $dist(p_{k_n}, p_{k_1}) \leq r$ and $|t_{k_n} - t_{k_1}| \geq \Delta t_s$, it means that P can be represented by the stay point $stop_k$, the corresponding virtual position of P. And θ_{times} is used to decide whether to add subsequent points p (satisfying $dist(p, p_{k_n}) \leq r$) to P. Reducing the value of θ_{times} can improve the accuracy of stay point detection.

The stay point is expressed by a quaternion $stop_k : \{lat_k, lng_k, arvt_k, levt_k\}$, where:

$$lat_k = \sum_{k_i=k_1}^{k_n} \frac{lat_{k_i}}{|P|}, k_1 \leq k_i \leq k_n \tag{1}$$

$$lng_k = \sum_{k_i=k_1}^{k_n} \frac{lng_{k_i}}{|P|}, k_1 \le k_i \le k_n \tag{2}$$

In the formula, lat_k represents the average latitude of set P and lng_k represents the average longitude of set P. The time of entering the stay point $arvt_k$ equals t_{k_1}. The time of leaving the stay point $levt_k$ equals t_{k_n}. And average velocity can be add to $stop_k$: $\{lat_k, lng_k, arvt_k, levt_k, v_k\}$ if needed.

Algorithm 1: Stay Point Detection Algorithm (nSPD)

Input: original trajectory T, threshold Δt_s, r and θ_{times}
Output: the corresponding stay point trajectory T^s

```
1      p = T.firstPoint();
2      while p != null
3          p' = p.nextPoint();
4          times = 0;
5          while p' != null
6              if ||p - p'||_gc <= r
7                  p.addStPt(p');
8                  p' = p'.nextPoint();
9              else if times < θ_times
10                 p = p.lastStP();
11                 times++;
12             else if |p.stPts()| > 0 and p.lastStP().t - p.t > Δt_s
13                 T^s.add((p.stPts()));
14                 break;
15             else
16                 T^s.add(p);
17                 p = p';
18                 break;
19     return T^s;
```

In the algorithm, $p.stPts()$ is the stay point when p is the pivot point, and $p.lastStP()$ is the last point in point p. When calculating the distance between points, we use the great circle distance [9]. The great circle distance from point p to p' is $||p - p'||_{gc}$. Considering that the region of stay points is small, we use the *haversine* formula to calculate the great circle distance.

3.2 Chasing Pattern Discovery Algorithm Based on Stay Point Detection

Given the candidate original trajectory set $\mathbb{T}c$, if a given original trajectory T ($T \notin \mathbb{T}c$) with the starting point p_1 is chased by an original trajectory T' T' ($T' \in \mathbb{T}c$), then T' should pass p_1. In fact, any point in T can be considered as a starting point. So we first align all the trajectories in $\mathbb{T}c$ with T, intercepting the remaining part of the trajectories after p_1 (including point p_1), then convert the candidate original trajectories in $\mathbb{T}c$ into the stay point trajectories one by one, thus forming $\mathbb{T}c^s$, and convert T into T^s.

After that, we find which stay point trajectory T_i in $\mathbb{T}c^s$ is chasing T^s, according to the time difference tolerance Δtd, the average velocity threshold α, the duration parameter τ, as well as the number of the stay points and the distance between the stay points int T_i and T^s, thus forming $\mathbb{R}^s$. The specific steps of chasing pattern discovery algorithm (sChPD) based on stay point detection are shown in Algorithm 2.

In the algorithm, the constraint on the velocity (12 lines) is optional. $P_s.t_{end}$ is the end time of the chasing trajectory pair, and $P_s.t_{start}$ is the start time of the chasing trajectory pair.

Algorithm 2: Stay-point-based Chasing Pattern Discovery Algorithm (sChPD)

Input: candidate stay points trajectory set $\mathbb{T}c^s$, stay points trajectory T^s, threshold Δtd, α and τ

Output: the corresponding chasing stay points trajectory set $\mathbb{R}^s$

```
1     Initialize ℝˢ;
2     for Tᵢ in 𝕋cˢ
3         p₁ = Tˢ.firstPoint();
4         p₂ = Tᵢ.firstPoint();
5         while p₁ != null and p₂ != null
6             if p₁.arvt < p₂.arvt
7                 if ||p₁ - p₂||_gc ≤ ε and |p₁.arvt - p₂.arvt| < Δtd
8                     Pₛ.add(p₂);
9                 p₁ = p₁.nextStayPoint(), p₂ = p₂.nextStayPoint();
10            if (1 + α) ≥ |Tˢ.v/Tᵢ.v| ≥ (1 - α)
11                if Pₛ.t_end - Pₛ.t_start > τ
12                    ℝˢ.add(Tᵢ.t_id, Pₛ);
13            i++;
14    return ℝˢ;
```

---

4   Experimental Evaluation and Algorithm Analysis

In this section, the performance of the proposed stay points detection method and the algorithm of discovering chasing pattern is evaluated with real taxi dataset.

4.1   Settings

Dataset. We use the S-Taxi, trajectories of 109 taxis in a city in northeast China in one day. The size of the dataset is about 168 MB, including 780,000 GPS track sampling points, and its sampling time is 3–10 s after cutting and preprocessing. The points of each trajectory consist of the following information: timestamp, latitude, longitude, and speed of the vehicle when it passed the current position. In addition, we use time intervals to cut and divide the trajectory to facilitate pattern mining and discovery.

Implementations. All experiments are performed on a computer with Intel Core i5-4200 CPU with 8 GB RAM running windows 10, and all the methods are implemented in Python 3.6.

Benchmark Algorithm. Chasing pattern discovery algorithm based on original points.

4.2 Evaluation of Stay Points Detection Algorithm

In the stay points detection algorithm, the parameters that affect the number of stay points detections are mainly the value of the switch time θ_{times}, the time threshold Δt_s and the distance threshold r. In Fig. 4, stay points (SP) percent represents the number of trajectories with stay points as a percentage of the total trajectories.

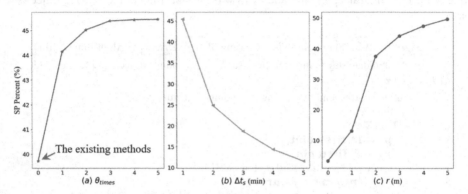

Fig. 4. Stay points (SP) percent on different parameters

The existing stop point detection methods do not use the switch time parameter, which can be considered as the case where the switch time θ_{times} is set to 0 in our proposed method. Regarding the switch time, it can be seen from Fig. 4(a) that there is a large difference between the case where the value of switch time is zero and non-zero, which means that the introduction of the switch time helps the discovery of stay points, and the increase in the value of switch time can be used to find more stay points in the trajectory. But limited by the dataset and other constraints, when the value of switch time is greater than 3, basically no more trajectories with stay points will be found. As for the time threshold, it can be seen from Fig. 4(b) that with the increase of the time threshold, the number of trajectories with stay points is found to decrease, which indicates that the increase of the time threshold effectively filters the set of stay points with small time intervals. However, if the threshold is too large, the normal stay points can also be filtered out. As for the distance threshold, it can be seen from Fig. 4(c) that the increase of the distance threshold can relax the conditions of stay points detection, so that more stay points can be detected. But as the distance threshold increases, those points that are not actually stay points are gathered in a set of stay points, which can cause a certain error.

4.3 Evaluation of Chasing Pattern Discovery Algorithm

Running Time. Figure 5 shows the running time of the two chasing pattern discovery algorithm in different time intervals. These two algorithms are based on the stay points and the original points. The time interval in the figure represents the intervals separated by the time of a day according to the given interval threshold, which is used to divide the trajectory. For example, the interval length in Fig. 5 is set to 30 min, then the trajectory

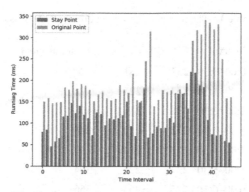

Fig. 5. Running time on different time interval

data of a day will be divided into sub-trajectory according to time and fall into 48 intervals (24 h have 48 time intervals with 30 min length).

In Fig. 5, we set distance threshold $\Delta d = 3$ m, time threshold $\Delta td = 2$ min, average rate threshold $\alpha = 0.5$, and duration $\tau = 5$ min. It can be seen that the chasing pattern discovery algorithm based on the stay points is more efficient in time running than the chasing pattern discovery algorithm based on the original point. This shows that the setting of the stay points shortens the time of finding trajectory with the chasing pattern while reducing the number of trajectory points.

The Number of Chasing Pattern Trajectories(CPT). In Table 1, we selected the trajectory data for 10 time intervals as the candidate trajectory set. The interval length of the divided trajectory is set to 30 min, and other parameter settings are the same as Fig. 5. Then we search the chasing pattern trajectories corresponding to each trajectory from the candidate set. Here we define the statistical method of the number of chasing pattern trajectories as follows. For each trajectory T_i in the given time interval, as long as there is another trajectory T_j that satisfies the condition of chasing T_i, then the total number of chasing pattern trajectories in the given time interval increases by 1. It can be seen from Table 1 that there are more chasing pattern trajectories found by the stay-points-based algorithm (CPT-SP) than that found by the original-points-based algorithm (CPT-OP) in most time intervals.

In order to find the effect of the interval length of the divided trajectory on the chasing pattern discovery, we set the interval length of the divided trajectory to 15 min, 30 min, and 60 min for experimental verification. The results are shown in Table 2. We know that as the interval length increases, the trajectory in each interval will contain more points, but the total number of trajectories will decrease. So the percent of CPT found is increasing along with the interval length increasing. From Table 2, we find that at each kind of interval length, the algorithm based on stay points always finds more trajectories of the chasing pattern than the algorithm based on original points in general.

Parameter Evaluation. Through experiments we find that the main parameters that affect the chasing pattern discovery algorithm are the distance threshold Δd and the chasing duration τ. We chose a 30-min time interval as the experimental data, which contains 324 trajectories. Except for the control parameters, other parameters are set the

Table 1. The number of CPT on different intervals

Interval	CPT-SP	CPT-OP	Total trajectory
0:00–0:30	**237**	231	324
1:00–1:30	**233**	225	312
2:00–2:30	**235**	231	308
3:00–3:30	**222**	218	280
4:00–4:30	**225**	214	296
5:00–5:30	**227**	221	304
6:00–6:30	**275**	265	356
7:00–7:30	**274**	270	364
8:00–8:30	279	279	376
9:00–8:30	285	287	380

Table 2. The percent of CPT on different interval length

Interval Length	CPT-SP Percent	CPT-OP Percent	Total Trajectory
15	0.72959	0.72325	32972
30	0.75863	0.74833	16796
60	0.78155	0.77608	7132

same as Fig. 5. As for the distance threshold parameter, it can be seen from Fig. 6(a) that increasing the distance threshold will find more chasing pattern trajectories. However, this does not mean that the larger the distance threshold is, the better the algorithm will perform, because for two trajectories that are far away and there is no chasing pattern between them, they will be misjudged as having a chasing pattern, which will cause errors. As the distance threshold increases, the differences between the original-points-based method and the stay-points-based methods will reduce. In extreme cases the original-points-based method even find more chasing pattern trajectories.

As for the duration, it can be seen from Fig. 6(b) that the increase in the duration of the chasing pattern trajectory will reduce the number of chasing pattern trajectories found. This shows that the duration of the trajectory with the chasing pattern will not be very long due to the limitation of other parameters such as the distance threshold, and increasing the duration will filter out some trajectories that have been found. In addition, when the duration τ is relatively short, the method based on the original points will have more semantic information than the method based on the stay points, so more chasing pattern trajectories will be found.

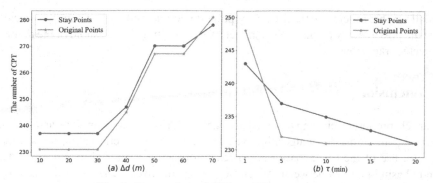

Fig. 6. The number of CPT on different parameter

5 Related Work

There is related work about chasing pattern discovery based on trajectory data. The trajectory data is a set of spatio-temporal sequences which chasing pattern is essentially a kind of sequential pattern or frequent pattern. Agrawal R and Srikant R first proposed sequential pattern mining in 1995 [10], and subsequent related research focused on spatial information only, rarely considering temporal information [11, 12]. Related work that considered temporal features mostly focused on the mining of periodic patterns [13, 14], which is not applicable to aperiodic chasing patterns. Some other special behavioral patterns, such as gathering, convoy, swarm, had received attention and research [15–17]. This kind of work does not involve the discovery of chasing patterns, but is helpful to the research of chasing patterns discovery in trajectory data.

Chasing pattern is a behavior pattern involving a pair of moving objects' trajectories. It has both coarse-grained and fine-grained spatio-temporal characteristics. Existing chasing pattern matching methods [4, 5] only involved fine-grained spatio-temporal features. If they are used for moving objects with a large range of activities, coarse-grained spatio-temporal features often cannot be found. In this article, stay-points-based chasing pattern discovery methods can process a larger range of trajectory data, and consider the fine-grained behavior characteristics of moving objects at the same time.

Existing stay point detection technologies aim to find frequent stay information at the same place at the same time, which is used to predict events such as refueling, taxiing, and switching classes. In the prediction of refueling events, in order to reduce the driver's waiting time when refueling, the stay point detection technology focuses on finding the stay position and time in the trajectory in order to optimize the distribution of gas stations [18]. In the prediction of taxi events, the stay point detection technology focuses on calculating the waiting time based on the arrival and departure times of the stay points [19]. In the prediction of the exchange class event, the stay point detection technology is to find a specific time period based on the change in the speed of the vehicle. So as to divide the driving trajectory of the same taxi to different drivers [7]. And other related studies are to find points of interest by the number of stay points in order to plan tourist routes [20]. The common feature of these stay point detection technologies is that the time and space range of the stay points are relatively fixed, and the stay time in the chasing pattern may be long or short, and the space range may be large or small. In this article, we design a stay point detection method that can dynamically detect the possible stay points in each trajectory by adjusting the threshold of the number of detected transitions.

6 Conclusion

This article proposes a chasing pattern discovery model based on stay points, and designs a chasing pattern discovery algorithm based on stay point detection algorithms. It can make full use of the time, space, and speed information contained in trajectory data to find chasing patterns between trajectories more effectively. The experiments verify the effectiveness and performance of the chasing pattern discovery algorithm proposed in this article. However, the parameters in the model designed in this article require a large number of experiments and data analysis to determine the appropriate values for the specific data set. The future work should use machine learning methods to obtain appropriate parameters automatically through a large amount of data training as well as abductive learning [21] methods with the stay point detection technology to find other behavior patterns in trajectory data. In addition, more practical application schemes of behavior pattern discovery methods should be discussed.

References

1. Alvares, L.O., Loy, A.M., Renso, C., Bogorny, V.: An algorithm to identify avoidance behavior in moving object trajectories. J. Braz. Comput. Soc. **17**(3), 193–203 (2011). https://doi.org/10.1007/s13173-011-0037-3
2. Zhao, X.-L., Xu, W.-X.: A clustering-based approach for discovering interesting places in a single trajectory. In: ICICTA, pp. 429–432 (2009)
3. Siqueira, F.D.L., Bogorny, V.: Discovering chasing behavior in moving object trajectories. Trans. GIS **15**(5), 667–688 (2011)
4. Kang, C.H., Hwang, J.R., Li, K.J.: Trajectory analysis for soccer players. In: ICDMW, pp. 377–381 (2006)
5. Zheng, Y., Zhou, X.: Computing with Spatial Trajectories, pp. 179–196. Springer, New York (2011). https://doi.org/10.1007/978-1-4614-1629-6
6. Alvares L. O., Bogorny V., Kuijpers B.: Towards semantic trajectory knowledge discovery. Data Min. Knowl. Discov., 1–12 (2007)
7. Zhang, D., Sun, L., Li, B.: Understanding taxi service strategies from taxi GPS traces. IEEE Trans. Intell. Transp. Syst. **16**(1), 123–135 (2015)
8. Zheng, Y., Zhang, L., Ma, Z., et al.: Recommending friends and locations based on individual location history. ACM Trans. Web **5**(1), 1–44 (2011)
9. Gade, K.: A non-singular horizontal position representation. J. Navig. **63**(3), 395–417 (2010)
10. Agrawal, R., Srikant, R.: Mining sequential patterns. In: ICDE, pp. 3–14 (1995)
11. Zhang, C., Han, J., Shou, L., et al.: Splitter: mining fine-grained sequential patterns in semantic trajectories. PVLDB **7**(9), 769–780 (2014)
12. Zhang, C., Ma, X., Zheng, Y., et al.: Assembler: efficient discovery of spatial co-evolving patterns in massive geo-sensory data. In: KDD, pp. 1415–1424 (2015)
13. Li, Z., Wang, J., Han, J.: ePeriodicity: mining event periodicity from incomplete observations. IEEE Trans. Knowl. Discov. Data Eng. **27**(5), 1219–1232 (2015)

14. Jindal, T., Giridhar, P., Tang, L.-A., et al.: Spatiotemporal periodical pattern mining in traffic data. In: SIGKDD Workshop on Urban Computing, pp. 1–8 (2013)
15. Zheng, K., Zheng, Y., Yuan, N.J., et al.: Online discovery of gathering patterns over trajectories. IEEE Trans. Knowl. Discov. Data Eng. **26**(8), 1974–1988 (2014)
16. Jeung, H., Yiu, M.L., Zhou, X., et al.: Discovery of convoys in trajectory databases. PVLDB, 1068–1080 (2008)
17. Li, Z., Ding, B., Han, J., et al.: Swarm: mining relaxed temporal moving object clusters. PVLDB, 723–734 (2010)
18. Zhang, F., Yuan, N.J., Wilkie, D., et al.: Sensing the pulse of urban refueling behavior: a perspective from taxi mobility. ACM Trans. Intell. Syst. Technol. **6**(3), 1–23 (2015)
19. Qi, G., Pan, G., Li, S., et al.: How long a passenger waits for a vacant taxi—large-scale taxi trace mining for smart cities. In: Green Computing and Communications, pp. 1029–1036 (2013)
20. Yuan, J., Zheng, Y., Xie, X., et al.: T-Drive: enhancing driving directions with taxi drivers' intelligence. IEEE Trans. Knowl. Discov. Data Eng. **25**(1), 220–232 (2013)
21. Zhou, Z.-H.: Abductive learning: towards bridging machine learning and logical reasoning. Sci. China Inf. Sci. **62**(7), 1–3 (2019)

A Cloud-Based Platform for ECG Monitoring and Early Warning Using Big Data and Artificial Intelligence Technologies

Chunjie Zhou$^{(\boxtimes)}$, Ali Li, Zhiwang Zhang, Zhenxing Zhang, and Haiping Qu

Department of Information and Electrical Engineering, Ludong University, Shandong, China
lucyzcj@ldu.edu.cn

Abstract. The prevalence of heart failure is increasing and is among the most costly diseases to society. Early detection of heart disease would provide the means to test lifestyle and pharmacologic interventions that may slow disease progression. However, the massive medical data have the following characteristics: real-time, high frequency, multi-source, heterogeneous, complex, random and personality. All of these factors make it very difficult to detect heart disease timely and make heart-warning signals accurately. So big data and artificial intelligence technologies are introduced to the field of health care, in order to discover all kinds of diseases and syndromes, and excavate valuable information to provide systematic decision-making for the diagnosis and treatment of heart. A cloud-based platform for ECG monitoring and early warning - HeartCarer is created, including a personalized data description model, the evaluation strategy of physiological indexes, and warning methods of trend-similarity about data flow. The proposed platform is particularly appropriate to address the early detection and warning of heart, which can provide users with efficient, intelligent and personalized services.

Keywords: Heart failure · Massive medical data · Big data · Early detection and warning · Cloud-based platform

1 Introduction

Heart failure (HF) prevalence is increasing and is among the most costly diseases to Medicare [1–3]. HF affects approximately 5.7 million people in the United States, and about 825,000 new cases per year with 33 billion total annual cost [4,5]. The lifetime risk of developing HF is 20% at 40 years of age [4,6,7]. HF has a high mortality

C. Zhou—This research was partially supported by the Project of Shandong Province Higher Educational Science and Technology Program (No. J12LN05); the grants from the National Natural Science Foundation of China (No. 61202111, 61273152, 61303017); the Project Development Plan of Science and Technology of Yantai City (No. 2013ZH092); the Doctoral Foundation of Ludong University (No. LY2012023); the US National Library of Medicine (No. R01LM009239); the Natural Science Foundation of Shandong Province China (No. ZR2011GQ001); and Scientific Research Foundation for Returned Scholars of Ministry of Education of China (43th).

Y. Nah et al. (Eds.): DASFAA 2020 Workshops, LNCS 12115, pp. 60–72, 2020.
https://doi.org/10.1007/978-3-030-59413-8_5

rate: 50% within 5 years of diagnosis [8,9] and causes or contributes to approximately 280,000 deaths every year [10,11]. Approximately 20–30% patients are readmitted after 30 days and nearly 50% are readmitted within the next six months. Moreover, with the aging of the population, this tendency will continue to increase [12–14]. There has been relatively little progress in slowing progression of HF severity largely because there are no effective means of early detection of HF to test interventions. Early detection of heart disease would provide the means to test lifestyle and pharmacologic interventions that may slow disease progression.

However, the massive medical data have the following characteristics: real-time, high frequency, multi-source, heterogeneous, complex, random and personality. Most of the data are collected from sensors, video, cameras etc. at the low level. The result for processing systems is a very diverse collection of different types and formats of data. Processing and aggregation of these data is a major challenge especially when analyzing in real-time large streams of physiological data such as electrograph (EEG) and electrograph (ECG). All of these factors make it very difficult to detect heart disease timely and make heart-warning signals accurately.

So big data and artificial intelligence technologies are introduced to the field of health care, in order to discover all kinds of diseases and syndromes, and excavate valuable information to provide systematic decision-making for the diagnosis and treatment of heart. In this paper, a cloud-based platform for ECG monitoring and early warning - HeartCarer is created based on big data and artificial intelligence technologies.

This research topic comes from a real research project. In this study, data resulting from the HeartCarer telemonitoring study was employed. The records are composed of biosignals collected on a daily basis, in particular, before the occurrence of one event, together with its prediction (decompensation HF and normal condition). The proposed platform is particularly appropriate to address the early detection and warning of heart, which can provide users with efficient, intelligent and personalized services.

The main contributions of this paper are summarized as follows:

- We proposed a personalized data description model;
- We proposed the evaluation strategy of physiological indexes, and warning methods of trend-similarity about data flow;
- A cloud-based platform for ECG monitoring and early warning is created.

2   Related Work

The goal of prediction is to analyze the relationships among existing data and detect abnormalities, so as to make corresponding early warnings. Computer-assisted methods, such as medical data mining or medical knowledge discovery, are effective tools for predicting disease and are used to explore hidden relationships among massive data sets.

Karaolis [17] developed a data mining system for assessing heart-related risk factors using association analysis based on the Apriori algorithm. The results show that smoking is one of the major risk factors that directly affect coronary heart disease. The literature [18] uses Doppler radar to monitor the heartbeat and respiration of the elderly at night and excludes the abnormality of the elderly's body motion. The monitoring results are used to monitor the occurrence of respiratory disorders in the elderly during sleep. By monitoring the physical parameters of electrocardiogram, blood oxygen saturation, blood pressure, and weight of the elderly, combined with daily activities and walking, liturature [19] determines whether there is chronicity heart failure threshold by judging each elderly's individualized threshold. Karlberg and Elo [20] analyzed data on ischemic heart disease (IHD) and coronary risk factors in the population. Calculations show that the age-specific rate of acute myocardial infarction (AMI) is only slightly different, and the incidence of angina is 3 to 10 times higher in people older than 50 years. Kunc [21] presented a simulation result that can be used to assess coronary heart disease, congestive heart failure, and end-stage renal disease in Slovenia. CoCaMAAL is a middleware-centric, cloud-based solution [16] that proposes different methods for handling various types of data such as vital signs, activity logs, and location logs. However, these systems are often only suitable for a certain application point, only consider a single context, and the efficiency of real-time processing is not high.

The assessment of similarity between time series is a central concept in knowledge discovery and generally consists of evaluating the similitude between two different time series. Two main groups of algorithms can be identified: time domain and transform-based methods. The former work directly with the raw signals (eventually with some preprocessing) and the main goal is to derive a measure (scalar) based on the comparison of the original time series. Euclidean distance for signals with the same length and dynamic time warping technique for signals with different lengths, are well-known examples of such algorithms [22]. Due to the high dimensionality of time series, most of the approaches perform dimension reduction on original data (transformed-based methods). This second group includes, among others, discrete Fourier transform [23] and singular value decomposition [24]. Other authors used the principal component analysis (or Karhunen Loeve transform) [23], while others applied methods based on discrete wavelet transform [25].

In summary, the existing research work has various deficiencies and cannot meet the needs of early detection and prediction of heart failure. Therefore, this paper proposes a Bayesian-based personalized event description model, a data stream trend similarity assessment strategy based on time series. In addition, based on the open prediction and early warning platform for heart failure, the proposed theory and method are validated.

3   System Overview and Methodology

3.1   System Architecture

A context aware cloud-based platform - HeartCarer is proposed here for ECG monitoring and early warning. This platform makes it possible for third-party systems to provide high level monitoring data and obtain detection and prediction services from the system. User can work with the system to define their preferences regarding the input and output of the system. The proposed platform architecture is shown in Fig. 1.

Researchers and engineers working with real-time signals perform similar preprocessing and processing steps prior to making inferences. The collected data can be used both in real-time and off-line to derive multiple inferences about the patients condition. However, applications in the healthcare domain are fairly limited due to the

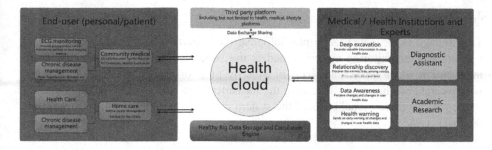

Fig. 1. The framework of the platform

processing and network demands on the supporting infrastructure. A real-world health-care application requires analyzing high-resolution sensor data in real time as well as data from other sources simultaneously, for many users at the same time. Processing the whole data on single machine locally is not practical due to computational limitations, reliability, scalability, failure/recovery and power consumption concerns.

To address this issue of application dependent implementations there is a need for processing platforms that are efficient enough to operate under real-world hardware and software constraints. Also, at the same time to be general enough to support different problems and applications. This work suggests an architecture that tries to solve this problem using intelligent distribution of the computational load using a publication/-subscription scheme. In this regard, the projects goal is to design and test an architecture which can scale to handle very large number of users and can act as a platform for processing real-time health analytics/inferences.

Basically, telemonitoring involves the transfer of physiological data and symptoms from patients at home (anytime and anywhere) to health-care providers. This allows more frequent assessment of a patients HF status and earlier recognition of hemo-dynamic deterioration than would be possible in common clinical practice. In effect, remote monitoring allows professionals to play a proactive role in daily care by imple-menting more effective and personalized therapies.

The methodology is based on a trend similarity measure, followed by a predictive procedure. The similarity scheme combines the Haar wavelet decomposition, in which signals are represented as linear combinations of a set of orthogonal bases, with the Karhunan-Loeve transformation (KLT), which allows for the optimal reduction of that set of bases. The trend similarity measure is then indirectly calculated by means of the coefficients obtained in both time-series description. The prediction strategy assumes that trends of physiological data common to patients with similar disease progression may have prognostic value in the prediction process. Therefore, using an approach sim-ilar to the k-nearest neighbor (k-NN), an estimation of the biosignals future values is performed, supported on a set of similar time series previously identified in the histori-cal dataset.

3.2 Data Sample Collection

The strategy is evaluated using physiological data resulting from HeartCarer, which is a home telemonitoring system aimed at the supervision of HF patients, enabling intervention when appropriate. This is done by monitoring physiological body signs with wearable technology, processing the measured data and giving recommendations to the patient. Professional users of the system can use the measured data to give user feedback.

Table 1. Description of properties in the dataset

S. no	Properties	Description
1	Age	Age in years
2	Sex	Sex (1 = male; 0 = female)
3	Weight	Weight of the corresponding patient
4	Waist circumference	Male > 40 in. and female > 35 in.
5	Smoking	If yes = 1 and no = 0
6	CP Type	Chest pain type
7	Systolic BP	Systolic blood pressure in mm/Hg
8	Diastolic BP	Diastolic blood pressure in mm/Hg
9	Serum cholesterol	Serum cholesterol in mg/dl
10	Fasting blood sugar	Fasting blood sugar >120 mg/dl
11	Restecg	Resting electrocardiographic results
12	Thalach	Maximum heart rate achieved
13	Exang	Exercise induced angina
14	Old peak	ST depression induced by exercise relative to rest
15	Slope	The slope of the peak exercise ST segment
16	Ca	Number of major vessels (0–3) colored by fluoroscopy
17	Thal	3 = normal; 6 = fixed defect; 7 = reversible defect

This system was used in a clinical observational study carried out with 168 patients from six clinical centers in China. The trial had an enrolment phase of 9 months with 12 months of patient follow up. During the clinical study, patients were requested to daily measure (during the morning period approximately at the same hour), weight, blood pressure, and, using a vest, the heart rate, and bioimpedance. Also the heart rate, respiration rate, and activity were monitored during the night by means of a bed sensor. Moreover, they were requested to complete two questionnaires of symptoms and mood/-general well-being each day. The measured properties are shown in Table 1. From the 168 patients recruited, 132 (78%) were considered analyzable, that is, with more than 30 days of telemonitoring measurements. Additionally, HF related events were recorded. Six cardiologists have analyzed the data, identifying which patients had experienced a decompensation event requiring hospitalization (47 patients) and which patients had not (85 patients).

The obtained results suggest, in general, that the physiological data have predictive value, and in particular, that the proposed scheme is particularly appropriate to address the early detection of HF decompensation.

4   Personalized Data Description Model

Each time a patient is admitted to the hospital the reason for the admission, as determined by the medical practitioner in primary charge during the admission, is recorded in the patients medical history. This is performed using a standardized international 'codebook', the International Statistical Classification of Diseases and Related Health Problems (ICD) [31], a medical classification list of diseases, injuries, symptoms, examinations, physical, mental or social circumstances issued by the World Health Organization. The ICD has a tree-like structure; at the top-most level codes are grouped into 12 chapters, each chapter encompassing a spectrum of related health issues. A disease progression is seen as being reflected by a patients admission history $H = a_1 \rightarrow a_2 \rightarrow ... \rightarrow a_n$ where a_i is a discrete variable whose value is an ICD code corresponding to the ith of n admissions on the patients record. The parameters of the underlying firstorder Markov model are then learnt by estimating transition probabilities $p(a' \rightarrow a'')$ for all transitions encountered in training (the remaining transition probabilities are usually set to some low value rather than 0). The model can be applied to predict the admission a_{n+1} expected to follow from the current history by likelihood maximization:

$$a_{n+1} = \arg \max_a p(a_n \rightarrow a) \tag{1}$$

Alternatively, it may be used to estimate the probability of a particular diagnosis a^* at some point in future:

$$p_f(a^*) = \sum_a [p(a \rightarrow a^*)p_f(a)] \tag{2}$$

or to sample the space of possible histories:

$$H' = a_1 \rightarrow a_2 \rightarrow ... \rightarrow a_n \dashrightarrow a_{n+1} \dashrightarrow a_{n+2}... \tag{3}$$

Our aim is to predict the probability of a specific admission a following the patient history H:

$$p(H \rightarrow a|H) \tag{4}$$

A history H is represented using a history vector $v = v(H)$ which is a fixed length vector with binary values. Each vector element corresponds to a specific admission code (except for one special element explained shortly) and its value is 1 if and only if the corresponding admission is present in the history:

$$\forall a \in A . v(H)_{i(a)} = \begin{cases} 1 : \exists j . H = H_1 \rightarrow a_j \rightarrow H_2 \wedge a = a_j \\ 0 : otherwise \end{cases} \tag{5}$$

where A is the set of admission codes, $i(a)$ indexes the admission code a in a history vector, and $H_{1,2}$ may take on degenerate forms of empty histories.

The disease progression modelling problem at hand is thus reduced to the task of learning transition probabilities between different patient history vectors:

$$p(v(H) \rightarrow v(H^{'})) \tag{6}$$

It is important to observe that unlike in the case of Markov process models working on the admission level when the number of possible transition probabilities is close to n_a^2, here the transition space is far sparser. Specifically, note that it is impossible to observe a transition from a history vector which codes for the existence of a particular past admission to one which does not, that is:

$$v(H)_{i(a)} = 1 \wedge v(H^{'})_{i(a)} = 0 \Rightarrow p(v(H) \rightarrow v(H^{'})) = 0 \tag{7}$$

5   Data Flow Trend Similarity Measure Method

The correlation between a patient p_j and a data e_0 is the log function of the probability of abnormality in physiological index e_0 divided by the probability of no abnormality in e_0.

$$corr(e_0, p_j) = log \frac{Pr(F_0 | F_{j,1}, F_{j,2}, ..., F_{j,k})}{Pr(F_0^c | F_{j,1}, F_{j,2}, ..., F_{j,k})} \tag{8}$$

$corr(e_0, p_j)$ indicates the relationship between p_j and e_0. The higher $corr(e_0, p_j)$ is, the more relevant e_0 is to patient j, and with the greater probability that patient j is abnormal in data e_0. We can simplify the calculation of $corr(e_0, p_j)$ as follows.

$$corr(e_0, p_j) = log \frac{Pr(F_0)}{Pr(F_0^c)} + \sum_{i=1}^{k} \frac{logPr(F_{j,i} | F_0)}{Pr(F_{j,i} | F_0^c)} \tag{9}$$

To future simply the above formula, we make use of the characteristic of our data. Since a data has only two states normal or abnormal, i.e. F_0 or F_0^c, using the property of probability, $Pr(F_0^c)$ and $Pr(F_{j,i} | F_0^c)$ in (9) can be eliminated in $corr(e_0, p_j)$, as shown below.

$$log \frac{Pr(F_0)}{1 - Pr(F_0)} + \sum_{i=1}^{k} log \frac{Pr(F_{j,i} | F_0)(1 - Pr(F_0))}{Pr(F_{j,i} - Pr(F_{j,i} | F_0^c)Pr(F_0))} \tag{10}$$

From the concept of conditional probability, we know that a joint probability is

$$Pr(F_{j,i} | F_0) = Pr(F_{j,i} | F_0)Pr(F_0) \tag{11}$$

Then we can further simplify

$$corr(e_0, p_j) = (k-1)log \frac{1-\alpha}{\alpha} + \sum_{i=1}^{k} log \frac{\beta_{j,i}}{\gamma_{j,i} - \beta_{j,i}} \tag{12}$$

where

$$\alpha = Pr(F_0) = \frac{number\,of\,exceptional\,events}{number\,of\,events} \tag{13}$$

$$\gamma_{j,i} = Pr(F_{j,i}) = \frac{number\,of\,events\,that\,F_{j,i}\,holds}{number\,of\,events} \tag{14}$$

$$\beta_{j,i} = Pr(F_{j,i}|F_0) = \frac{Pr(F_{j,i},F_0)}{Pr(F_0)} = \frac{1}{\alpha}\frac{number\,of\,events\,that\,both\,F_0\,and\,F_{j,i}\,holds}{number\,of\,events} \tag{15}$$

We also apply smoothing techniques for the above formula. The next step is ranking $corr(e_0, p_j)$ with respect to all the data in descending order. The higher a data in the list, the more possible that p_j is abnormal in related to this data.

6   System Demonstration

The basic vital sign data in this system comes from the cooperative medical institutions. The real-time ECG data is collected using the 12-lead ECG monitoring equipment-HeartView 12BT- from Israeli Aerotel Medical System. This device supports two transmission modes of Bluetooth and voice. The transmission mode of Bluetooth means that the HV device can transmit the detected ECG data to the external Bluetooth media in real time through Bluetooth. The voice transmission mode refers to that the device can convert the ECG data into sound signals for transmission to the remote ECG monitoring center. Then the ECG center will restore the ECG based on the received voice signals. In this system, we use Bluetooth as the transmission channel, and send the ECG data to a corresponding APP first. From the APP to the ECG monitoring cloud center, the ECG interpretation and screening are performed by the automatic interpretation system.

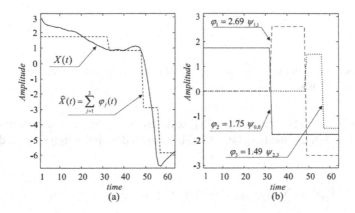

Fig. 2. Signal approximation using the Haar WaveletKLT decomposition scheme. (a) Actual signal and approximation (dashed line). (b) Specific wavelet bases used in the approximation.

As can be seen in Fig. 2, the main characteristics of the signal are captured using only three bases. Although it is possible to obtain small approximation errors (by increasing the threshold ε, and thus, the number of bases), it should be noted that this is not the main focus of the proposed scheme. Effectively, the final goal is to identify the main characteristics of the signal, that is, the main trends or behavior.

In this particular case, the most important basis is $\varphi_1(t) = 2.69\psi_{1,1}$ (largest coefficient). This means that the highest variation in the signal occurs between the instants, as can be confirmed by inspecting Fig. 2(a). Moreover, the second basis, $\varphi_2(t) = 1.75\psi_{0,0}$, is the one that reflects the contribution of the complete signal. Since the coefficient has a positive value, it can be concluded that the signal presents a global positive variation, that is, the mean of the first half, instants, is higher than the mean of the second half, instants. Finally, a similar conclusion can be drawn for the third basis, $\varphi_3(t) = 1.49\psi_{2,3}$. In fact, in the corresponding time region, instants, the signal presents a significant variation from a higher to a lower value, and thus, the coefficient has a positive sign.

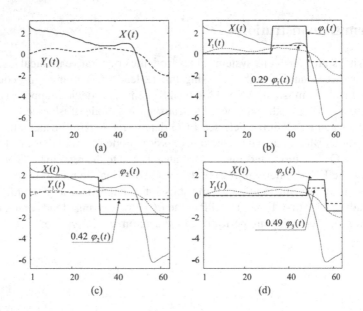

Fig. 3. Comparision of the two signals. (a) Signals to be compared, (b) first basis, (c) second basis, and (d) third basis.

The first signal to be compared, $Y_1(t)$, is described as (16) and shown in Fig. 3. In the second, Fig. 4, the same template, $X(t)$, is compared with the signal $Y_2(t)$, described as (17).

$$\hat{Y}_1(t) = \sum_{j=1}^{3} \alpha_j \varphi_j(t) = 0.29\varphi_1(t) + 0.42\varphi_2(t) + 0.49\varphi_3(t) \tag{16}$$

$$\hat{Y}_2(t) = \sum_{j=1}^{3} \alpha_j \varphi_j(t) = -0.93\varphi_1(t) - 0.49\varphi_2(t) + 0.21\varphi_3(t) \tag{17}$$

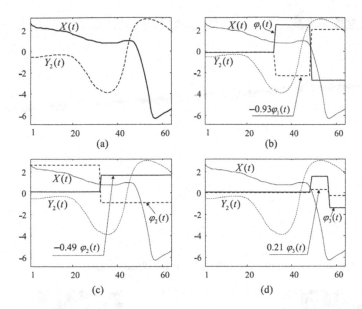

Fig. 4. Comparision of the two signals. (a) Signals to be compared, (b) first basis, (c) second basis, and (d) third basis.

In the first case, all the coefficients $\alpha_j (j = 1, 2, 3)$ are positive, thus having the same sign as the coefficients of the template (all equal to 1). From this simple statement, it can be concluded that template and signal present the same behavior, i.e., the same temporal trend (as can be observed in Fig. 3(a). As result the similarity measure, defined as (23), is $S_T(\Gamma, \Omega) = 1$. In the second example, it is observed that the first two coefficients are negative and the third is positive. Thus, it may be concluded that, in global terms, they do not present the same behavior (this can be observed for some intervals of the signal in Fig. 4(a). The similarity measure is in this case $S_T(\Gamma, \Omega) = 1/3$.

If the electrocardiogram is screened as invalid or abnormally abnormal electro-cardiogram, the patient will immediately receive the interpretation result of the auto-response from the electrocardiographic monitoring center. If the electrocardiogram is not clearly answered after the interpretation, it will automatically form a pending task for the supervising doctor or ECG experts, as shown in Fig. 5. The expert will do the manual interpretation and processing of the ECG, as shown in Fig. 6.

The system can help experts screen and process 80% of ECG and interpretation time. Meanwhile, it saves patients a lot of treatment and waiting time.

Fig. 5. The ECG monitoring cloud center

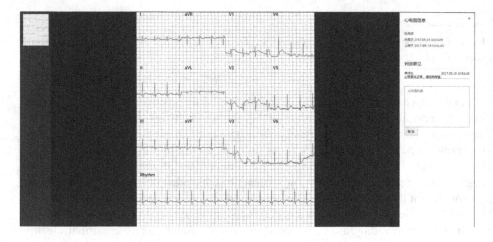

Fig. 6. The ECG

7   Conclusion

In conclusion, our study demonstrated that both unstructured and structured information can facilitate early detection of heart failure as early as two years prior to diagnosis. Through the introduction of big data and artificial intelligence technology, a cloud-based platform for ECG monitoring and early warning - HeartCarer is created. The proposed platform is particularly appropriate to address the early detection and warning of heart, which can effectively slow disease progression, save valuable rescue time for patients, reduce mortality, and provide users with efficient, intelligent and personalized services.

References

1. Mackay, J., Mensah, G.A., Mackay, J., et al.: The atlas of heart disease and stroke, **19**, 505–505 (2005)
2. Henriques, J., Carvalho, P.F., Rocha, T., et al.: Prediction of heart failure decompensation events by trend analysis of telemonitoring data. IEEE J. Biomed. Health Inform. **19**(5), 1757–1769 (2014)
3. Satija, U., Ramkumar, B., Manikandan, M.S.: Real-time signal quality-aware ECG telemetry system for IoT-based health care monitoring. IEEE Internet Things J. **PP**(99), 1 (2017)
4. Moss, T.J., Clark, M.T., Calland, J.F., et al.: Cardiorespiratory dynamics measured from continuous ECG monitoring improves detection of deterioration in acute care patients: a retrospective cohort study. PLoS ONE **12**(8), e0181448 (2017)
5. Hamidi, M., Ghassemian, H., Imani, M.: Classification of heart sound signal using curve fitting and fractal dimension. Biomed. Signal Process. Control **39**, 351–359 (2018)
6. Sandha, S.S., Kachuee, M., Darabi, S.: Complex event processing of health data in real-time to predict heart failure risk and stress (2017)
7. Satija, U., Ramkumar, B., Manikandan, M.S.: Automated ECG noise detection and classification system for unsupervised healthcare monitoring. IEEE J. Biomed. Health Inform. **22**(3), 722–732 (2018)
8. Sushmita, S., Newman, S., Marquardt, J., et al.: Population cost prediction on public healthcare datasets. In: International Conference on Digital Health. pp. 87–94. ACM (2015)
9. Eslamizadeh, G., Barati, R.: Heart murmur detection based on wavelet transformation and a synergy between artificial neural network and modified neighbor annealing methods. Artif. Intell. Med. **78**, 23–40 (2017)
10. Giamouzis, G., Agha, S.A., Ekundayo, O.J., et al.: Incident coronary revascularization and subsequent mortality in chronic heart failure: a propensity-matched study. Int. J. Cardiol. **140**(1), 55–59 (2010)
11. Hua, J., Zhang, H., Liu, J., et al.: Direct arrhythmia classification from compressive ECG signals in wearable health monitoring system. J. Circ. Syst. Comput. **27**(6), 1–13 (2018)
12. Giamouzis, G., Kalogeropoulos, A., Georgiopoulou, V., et al.: Hospitalization epidemic in patients with heart failure: risk factors, risk prediction, knowledge gaps, and future directions. J. Card. Fail. **17**(1), 54–75 (2011)
13. Wattal, S., Spear, S.K., Imtiaz, M.H., et al.: A polypyrrole-coated textile electrode and connector for wearable ECG monitoring. In: IEEE International Conference on Wearable and Implantable Body Sensor Networks, pp. 54–57. IEEE (2018)
14. Jekova, I., Krasteva, V., Leber, R., et al.: A real-time quality monitoring system for optimal recording of 12-lead resting ECG. Biomed. Signal Process. Control **34**, 126–133 (2017)
15. Cubo, J., Nieto, A., Pimentel, E.: A Cloud-based Internet of Things platform for ambient assisted living. Sensors **14**(8), 14070–14105 (2014)
16. Forkan, A., Khalil, I., Tari, Z.: CoCaMAAL: a cloud-oriented context-aware middleware in ambient assisted living. Future Gener. Comput. Syst. **35**(35), 114–127 (2014)
17. Karaolis, M.A., Moutiris, J.A., Hadjipanayi, D., et al.: Assessment of the risk factors of coronary heart events based on data mining with decision trees. IEEE Trans. Inf Technol. Biomed. **14**(3), 559–566 (2010)
18. Kagawa, M., Yoshida, Y., Kubota, M., et al.: An overnight vital signs monitoring system for elderly people using dual microwave radars. In: Microwave Conference Proceedings, pp. 590–593. IEEE (2012)
19. Fanucci, L., Saponara, S., Bacchillone, T., et al.: Sensing devices and sensor signal processing for remote monitoring of vital signs in CHF patients. IEEE Trans. Instrum. Meas. **62**(3), 553–569 (2013)

20. Lee, J.V., Chuah, Y.D., Chieng, K.T.H.: Smart elderly home monitoring system with an android phone. Int. J. Smart Home **7**(3), 670–678 (2013)
21. Atanasijevi-Kunc, M., Drinovec, J., Ruigaj, S., et al.: Simulation analysis of coronary heart disease, congestive heart failure and end-stage renal disease economic burden. Math. Comput. Simul. **82**(3), 494–507 (2012)
22. Liao, T.W.: Clustering of time series data-a survey. Pattern Recogn. **38**(11), 1857–1874 (2005)
23. Liu, C., Gao, R.: Multiscale entropy analysis of the differential RR interval time series signal and its application in detecting congestive heart failure. Entropy **19**(6), 251 (2017)
24. Yang, K., Shahabi, C.: A PCA-based similarity measure for multivariate time series. In: ACM International Workshop on Multimedia Databases, pp. 65–74. ACM (2004)
25. Chen, X., Yang, R., Ge, L., et al.: Heart rate variability analysis during hypnosis using wavelet transformation. Biomed. Signal Process. Control **31**, 1–5 (2017)
26. Ogiela, L., Tadeusiewicz, R., Ogiela, M.R.: Cognitive techniques in medical information systems. Comput. Biol. Med. **38**(4), 501–507 (2008)
27. Sahoo, P.K., Thakkar, H.K., Lee, M.Y.: A cardiac early warning system with multi channel SCG and ECG monitoring for mobile health. Sensors **17**(4), 711 (2017)
28. Fu, L., Li, F., Zhou, J., Wen, X., Yao, J., Shepherd, M.: Event prediction in healthcare analytics: beyond prediction accuracy. In: Cao, H., Li, J., Wang, R. (eds.) PAKDD 2016. LNCS (LNAI), vol. 9794, pp. 181–189. Springer, Cham (2016). https://doi.org/10.1007/978-3-319-42996-0_15
29. Frick, S., Uehlinger, D.E., Zuercher Zenklusen, R.M.: Medical futility: predicting outcome of intensive care unit patients by nurses and doctors-a prospective comparative study. Crit. Care Med. **31**(2), 456–461 (2003)
30. Wu, W.H., Bui, A.A., Batalin, M.A., et al.: MEDIC: medical embedded device for individualized care. Artif. Intell. Med. **42**(2), 137–152 (2008)
31. Kurian, J., Kurian, J., Huang, J.X., et al.: A Bayesian-based prediction model for personalized medical health care. In: IEEE International Conference on Bioinformatics and Biomedicine. IEEE Computer Society, pp. 1–4 (2012)

What Are MOOCs Learners' Concerns? Text Analysis of Reviews for Computer Science Courses

Xieling Chen[1], Di Zou[2(✉)], Haoran Xie[3], and Gary Cheng[1]

[1] Department of Mathematics and Information Technology,
The Education University of Hong Kong, Hong Kong, Hong Kong SAR
[2] Department of English Language Education,
The Education University of Hong Kong, Hong Kong, Hong Kong SAR
dizoudaisy@gmail.com
[3] Department of Computing and Decision Sciences, Lingnan University,
Hong Kong, Hong Kong SAR

Abstract. In MOOCs, course reviews are valuable sources for exploiting learn-ers' attitudes towards the courses provided. This study employed an innovative structural topic modeling technique to analyze 1920 reviews of 339 courses regard-ing computer science to understand what primary concerns the learners had. Nine major topics, including *course levels, learning perception, course assessment, teaching styles, problem solving, course content, course organization, critique,* and *learning tools and platforms* were revealed. In addition, we investigated how the identified nine topics varied across reviews with different ratings. Results indi-cated that negative reviews tended to relate more to issues such as *course assess-ment, learning tools and platforms,* and *critique,* while positive reviews concerned more about issues such as *course levels, course organization,* and *learning per-ception.* This study provided tutors with novel implications for developing online courses, particularly computer science courses.

Keywords: MOOCs · Online course reviews · Structural topic model

1 Introduction

Massive open online courses (MOOCs) possess significant potential to enable every learner to access to high-quality educational resources. Since the appearance of the first open online course in 2011, there has been an increasing interest in MOOCs worldwide [1–3]. With the continuous growth in the interest of MOOCs, many colleges have been encouraged to launch online courses. According to Shah (2018) [4], there have been 11,400 MOOCs offered by over 900 colleges worldwide till 2018, with an enrollment of about 101 million online learners. Practitioners and educators have been very interested in the determination of how successful the MOOCs they launched [5].

Compared with face-to-face interviews, it is more convenient to use MOOCs in terms of collecting and storing learners' comments, and the quality and usability of the data

© Springer Nature Switzerland AG 2020
Y. Nah et al. (Eds.): DASFAA 2020 Workshops, LNCS 12115, pp. 73–79, 2020.
https://doi.org/10.1007/978-3-030-59413-8_6

collected can be ensured. A number of previous studies have concerned about structured data analysis, for example, learners' clicks and grades [6, 7]. Nevertheless, unstructured textual data collected using interactive technologies, for example, discussion forums and online course reviews, remained to be explored [8, 9]. Course reviews offer an abundant source of information regarding learners' opinions towards issues such as course contents, instructors, and course platforms [10]. For example, Hew (2016) [11] conducted an inductive iterative coding method on comments of 965 participants in three top-rated MOOCs courses to uncover causes for their popularities. Five factors were revealed, including problem-centered learning, tutor accessibility and enthusiasm, active learning, peer interaction, and supportive learning resources. However, the processes and findings by using the manual coding method heavily depend on the expertise of human coders [12]. Besides, subjective biases result in the difficulty in replicating the findings. In addition, the sample sizes are limited [13].

With the advance of computer-based text analysis methods, this study sought to explore reviews comments of computer science MOOCs courses by the use of topic modeling, an automatic approach that has become increasingly popular for analyzing large-scale textual data in recent years [14–17]. For example, Chen et al. (2020) [18] presented a thorough review of *the British Journal of Educational Technology* during the period 1971–2018, with the adoption of bibliometrics and topic models. Xie et al. [19] reviewed studies concerning personalized learning collected from the Web of Science by adopting a coding scheme proposed with the basis of the constructivism theories.

The goal of this study was to analyze 1920 reviews for 339 courses concerning computer science to understand what primary concerns the MOOCs learners had, with the adoption of an innovative structural topic modeling technique [20, 21]. In addition, we further investigated how the identified nine topics varied across reviews with different ratings.

2   Materials and Methods

The flow chart of data collection and analyses is described as Fig. 1, including data acquisition, data preparation, and structural topic modeling, as elaborated in the following sections.

Fig. 1. Flow chart of data collection and analyses

2.1 Data Acquisition and Preparation

The procedure of data acquisition and preparation was as follows. The online course reviews were collected using Class Central[1], one of the best-known MOOCs review sites worldwide. It had been used in a number of studies [22, 23]. We first selected MOOCs in the category of computer science, after removing duplicates as well as excluding those without review comments, 339 unique MOOCs remained. We then created a web crawler program to automatically crawled course metadata as well as corresponding review comments. Besides, we excluded reviews without providing useful information, for example, "A completed this course," "B is taking this course right now," and "C completed this course. Person found this review helpful." Finally, 1920 reviews for 339 courses in relation to computer science were used for topic modeling and analyses (Table 1).

Table 1. Statistics of the review dataset.

Items	Frequency	Percentage
Distribution of course grade (number of courses)		
Grade 1	20	5.90%
Grade 2	24	7.08%
Grade 3	45	13.27%
Grade 4	109	32.15%
Grade 5	141	41.59%
Distribution of course grade (number of reviews)		
Grade 1	177	9.22%
Grade 2	96	5.00%
Grade 3	94	4.90%
Grade 4	321	16.72%
Grade 5	1232	64.17%

2.2 STM Model Setup

The 1920 review comments were used for STM modeling. First, extract key terms from comments with the adoption of natural language processing. Second, unify terms with similar meanings. For example, organize, organized, and organizing were all unified as organize. Third, remove stop words (e.g., the, is, you, a), punctuations, and numbers.

---

[1] https://www.class-central.com/.

Fourth, remove meaningless words, for example, add, get, one, lot, let, something, some-one, sometimes, take, and anyone. Fifth, filter unimportant words using term frequency-inverse document frequencies (TF-IDF). We empirically excluded terms with a TF-IDF value of less than 0.05. We then ran several models, with the number of topics ranging from three to 25. Two domain experts independently carried out comparisons of the results of the 23 models, and a nine-topic model was selected as the final model.

3 Results

3.1 Topic Interpretation

Table 2 displays the nine-topic STM results, together with the representative terms, topic proportions, as well as suggested topic labels. The top four topics being the most frequently mentioned in review comments included *Course organization* (18.41%), *Course levels* (17.29%), *Learning perception* (13.42%), and *Problem solving* (12.00%). The other five topics were *Course content* (8.47%), *Learning tools and platforms* (8.31%), *Course assessment* (8.09%), *Critique* (7.17%), and *Teaching style* (6.83%).

Table 2. Topic summary.

Topic label	Topic proportion	Top words
Course levels	17.29%	Difficulty, medium, hour, spend, easy, hard, beginner
Learning perception	13.42%	Goodpretty, technical, introduction, clear, inspire, knowledge, overview
Course assessment	8.09%	Answer, unit, install, question, search, staff, figure, code, engine, wrong, final, solution, exam
Teaching style	6.83%	Didactic, interesante, informatics, explicate, explication, innovation, profound
Problem solving	12.00%	Programming, challenge, solve, assignment, algorithm, problem, note
Course content	8.47%	Regression, graphlab, analytics, classification, clustering, machine, neural
Course organization	18.41%	Creative, organize, fun, awesome, field, session, task
Critique	7.17%	Reversible, energy, computation, talk, free, poor, waste
Learning tools and platforms	8.31%	Git, github, watch, web, video, java, tool

3.2 Topic Distribution Across Negative and Positive Reviews

We further investigated how the identified nine topics varied across reviews with different ratings. As shown in Fig. 2, negative reviews tended more to be related to issues such as *Course assessment*, *Learning tools and platforms*, and *Critique*, while positive reviews concerned more about issues such as *Course levels*, *Course organization*, and *Learning perception*. In addition, there were two topics, that is, *Course content* and *Problem solving*, tended to be equally concerned for both negative and positive reviews.

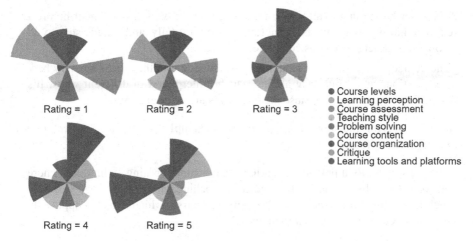

Fig. 2. Topic distributions across reviews with different ratings.

4 Discussion and Conclusion

This study analyzed 1920 reviews for 339 courses concerning computer science with the use of structural topic modeling, to understand what primary concerns the MOOCs learners had. We additionally investigated how the identified nine topics varied across reviews with different ratings. We empirically proved the effectiveness of the application of structural topic modeling of online course reviews. In particular, comments with high proportions to a particular topic were exactly relevant. For example, for the topic *Course levels*, examples of highly relevant review comments were as follows:

> "spending 5 h a week on it and found the course difficulty to be hard. good course but was hard i think that could also be Very hard."

> "spending 3 h a week on it and found the course difficulty to be easy. I made another dream come true - I've finished another Dr. Chuck's course! It appeared to be a lot better than I thought it to be. Really enjoyed it. It is useful and thought-provoking. Thank you, Dr. Chuck for all this!"

Second, for the topic *Learning perception*, examples of highly relevant review comments were as follows:

> "Excellent course. Cleanly separates the different pieces that go into making bitcoin, and explains each of them simply and clearly. Covers not only bit-coin and cryptocurrencies, but also the wider implications and applications of the blockchain to non-currency applications. Highly recommended for anyone interested in bitcoin or the blockchain."

Third, for the topic *Course assessment*, examples of highly relevant review comments were as follows:

"It's super buggy and totally frustrating. I got to the very last segment before I realized that my assignments did not even submit totally, and I made sure to go through the exact procedures completely."

"The course videos itself are moderately interesting, but the quizes are very bad. The questions are not clear at all and can be interpreted in different ways. It's frustrating to lose points because the questions are so confusing."

In addition, for the topic *Course content*, examples of highly relevant review comments were as follows:

"This was certainly a practical overview of machine learning techniques. There was very little discussion of the algorithms behind these techniques, certainly much less than even in Andrew Ng's Coursera course, which is itself supposedly fairly watered-down compared to many..."

This study provided tutors with novel insights into the design and improvement of online courses, particularly computer science courses. Regarding future research, there are three directions worth investigating. First, further investigation can be considered by defining negative and positive reviews [24] to explore which topics are more likely to appear in negative reviews. Second, the occurrence of these topics is also worth investigating. Besides, further investigation is encouraged to incorporate course metadata (e.g., course schedule and course duration) and learner metadata (e.g., course completion status) to explore factors affecting learners' satisfaction. In addition, future investigation using learning analytics to propose tools or systems for obtaining a better learning experience is also encouraged (e.g., [25]).

Acknowledgments. This research received grants from the Standing Committee on Language Education and Research (EDB(LE)/P&R/EL/175/2), the Education Bureau of the Hong Kong Special Administrative Region, the Internal Research Grant (RG93/2018-2019R), and the Internal Research Fund (RG 1/2019-2020R) of the Education University of Hong Kong.

References

1. Deng, R., Benckendorff, P., Gannaway, D.: Progress and new directions for teaching and learning in MOOCs. Comput. Educ. **129**, 48–60 (2019)
2. Gardner, J., Brooks, C.: Student success prediction in MOOCs. User Model. User-Adap. Inter. **28**, 127–203 (2018)
3. Zhu, M., Sari, A., Lee, M.M.: A systematic review of research methods and topics of the empirical MOOC literature (2014–2016). Internet High. Educ. **37**, 31–39 (2018)
4. Shah, D.: By the numbers: MOOCs in 2018 [Web log post] (2018)
5. Henderikx, M.A., Kreijns, K., Kalz, M.: Refining success and dropout in massive open online courses based on the intention–behavior gap. Dist. Educ. **38**, 353–368 (2017)
6. Sun, J.C.-Y., Kuo, C.-Y., Hou, H.-T., Lin, Y.-Y.: Exploring learners' sequential behavioral patterns, flow experience, and learning performance in an anti-phishing educational game. J. Educ. Technol. Soc. **20**, 45–60 (2017)

7. Tang, J.K.T., Xie, H., Wong, T.L.: A big data framework for early identification of dropout students in MOOC. In: Lam, J., Ng, K.K., Cheung, S.K.S., Wong, T.L., Li, K.C., Wang, F.L. (eds.) ICTE 2015. CCIS, vol. 559, pp. 127–132. Springer, Heidelberg (2015). https://doi.org/10.1007/978-3-662-48978-9_12

8. Chiu, T.K., Hew, T.K.: Factors influencing peer learning and performance in MOOC asynchronous online discussion forum. Australas. J. Educ. Technol. **34**, 16–28 (2018)

9. Wang, X., Yang, D., Wen, M., Koedinger, K., Rosé, C.P.: Investigating how student's cognitive behavior in MOOC discussion forums affect learning gains. In: International Educational Data Mining Society (2015)

10. Liu, S., Peng, X., Cheng, H.N., Liu, Z., Sun, J., Yang, C.: Unfolding sentimental and behavioral tendencies of learners' concerned topics from course reviews in a MOOC. J. Educ. Comput. Res. **57**, 670–696 (2019)

11. Hew, K.F.: Promoting engagement in online courses: what strategies can we learn from three highly rated MOOCS. Br. J. Educ. Technol. **47**, 320–341 (2016)

12. Zhou, L., Ye, S., Pearce, P.L., Wu, M.-Y.: Refreshing hotel satisfaction studies by reconfiguring customer review data. Int. J. Hospital. Manag. **38**, 1–10 (2014)

13. Mankad, S., Han, H.S., Goh, J., Gavirneni, S.: Understanding online hotel reviews through automated text analysis. Serv. Sci. **8**, 124–138 (2016)

14. Chen, X., Zou, D., Cheng, G., Xie, H.: Detecting latent topics and trends in educational technologies over four decades using structural topic modeling: a retrospective of all volumes of computer & education. Comput. Educ. **151**, 103855 (2020)

15. Chen, X., Yu, G., Cheng, G., Hao, T.: Research topics, author profiles, and collaboration networks in the top-ranked journal on educational technology over the past 40 years: a bibliometric analysis. J. Comput. Educ. **6**, 563–585 (2019)

16. Chen, X., Xie, H., Wang, F.L., Liu, Z., Xu, J., Hao, T.: A bibliometric analysis of natural language processing in medical research. BMC Med. Inform. Decis. Mak. **18**, 14 (2018)

17. Hao, T., Chen, X., Li, G., Yan, J.: A bibliometric analysis of text mining in medical research. Soft. Comput. **22**, 7875–7892 (2018)

18. Chen, X., Zou, D., Xie, H.: Fifty years of British Journal of Educational Technology: a topic modeling based bibliometric perspective. Br. J. Educ. Technol. **51**, 692–708 (2020)

19. Xie, H., Chu, H.-C., Hwang, G.-J., Wang, C.-C.: Trends and development in technology-enhanced adaptive/personalized learning: a systematic review of journal publications from 2007 to 2017. Comput. Educ. **140**, 103599 (2019)

20. Roberts, M.E., Stewart, B.M., Tingley, D.: stm: R package for structural topic models. J. Stat. Softw. **10**, 1–40 (2014)

21. Roberts, M.E., et al.: Structural topic models for open-ended survey responses. Am. J. Polit. Sci. **58**, 1064–1082 (2014)

22. Stratton, C., Grace, R.: Exploring linguistic diversity of MOOCs: implications for international development. Proc. Assoc. Inf. Sci. Technol. **53**, 1–10 (2016)

23. Kashyap, A., Nayak, A.: Different machine learning models to predict dropouts in MOOCs. In: 2018 International Conference on Advances in Computing, Communications and Informatics (ICACCI), pp. 80–85. IEEE (2018)

24. Hu, N., Zhang, T., Gao, B., Bose, I.: What do hotel customers complain about? Text analysis using structural topic model. Tour. Manage. **72**, 417–426 (2019)

25. Zou, D., Xie, H.: Personalized word-learning based on technique feature analysis and learning analytics. J. Educ. Technol. Soc. **21**, 233–244 (2018)

A New Context-Aware Method Based on Hybrid Ranking for Community-Oriented Lexical Simplification

Jiayin Song[1], Jingyue Hu[1], Leung-Pun Wong[2], Lap-Kei Lee[2], and Tianyong Hao[1]([✉])

[1] School of Computer Science, South China Normal University, Guangzhou, China
1015981339@qq.com, 974478195@qq.com, haoty@m.scnu.edu.cn
[2] School of Science and Technology, The Open University of Hong Kong, Hong Kong, China
{s1243151,lklee}@ouhk.edu.hk

Abstract. As a subtask of lexical substitution, lexical simplification aims to provide simplified words with the same semantic meaning in context. Focusing on the task, this paper proposes a new method based on a hybrid ranking strategy. The method consists of three parts including 1) substitution generation leveraging semantic dictionaries, 2) substitution selection utilizing part-of-speech tagging and word stemming, and 3) substitution ranking based on a hybrid approach. Through the evaluation on standard datasets, our method outperforms state-of-the-art baselines including Word2vec and Four-step method, indicating its effectiveness in lexical simplification.

Keywords: Lexical simplification · Word2vec · Ranking strategy · Lexical substitution

1 Introduction

To the readers of any document (e.g., article), unfamiliar or complex words are a significant barrier for text understanding. Text simplification refers to the process of converting a complex text into one with sentences of the same meaning, which would however be read and understood better by more people [1]. It includes several subtasks such as the identification of complex words and sentences, word sense disambiguation, lexical simplification and syntactic simplification. Lexical simplification, as a fundamental subtask, aims at replacing difficult words in a text with synonyms that are easier to understand by the reader [2]. For example, given a sentence "They locate food by smell, using sensors in the tip of their snout, and regularly feast on ants and termites," simplifying the target word "snout" into "nose" helps the understanding of the sentence meaning. Lexical simplification would be potentially useful for many Natural Language Processing (NLP) tasks, such as question answering, summarization, sentence completion, text simplification and text shortening. Particularly, it can potentially benefit second language learning [3].

© Springer Nature Switzerland AG 2020
Y. Nah et al. (Eds.): DASFAA 2020 Workshops, LNCS 12115, pp. 80–92, 2020.
https://doi.org/10.1007/978-3-030-59413-8_7

Typically, the task of lexical simplification obtains substitutes by picking synonyms from lexical resources as candidates, and then rank the candidates based on appropriateness in context. For example, some lexical substitution systems used word sense similarity directly to score substitute candidates [4], used vector space models approach [5, 6] to perform substitute ranking in context based on vector representations, and used context and target word embedding approach [7] to measure the similarity between representation vectors of target word embeddings and context embeddings.

The task of community-oriented lexical simplification is somewhat different from the normal lexical simplification task. The specified communities have their own list of words as required candidate words for simplification. For example, The Hong Kong Education Bureau requires transforming the complex English literature to be compatible with the word list defined by Hong Kong Education Bureau (EDB List), which includes approximately 4,000 words starting with the beginning of the letter A to the beginning of Z that all students in Hong Kong are expected to know upon graduation from primary school [8]. Another key problem is whether to preserve the original sentence grammatically. Existing methods have reported ranking algorithms utilizing "Continuous Bag-of-Words", n-grams, syntactic structures, and classifiers, e.g., [9, 10]. However, how to utilize semantic and context information together remains a challenging research topic for improving the ranking performance of substitution candidates.

This paper proposes a model for lexical simplification, which takes both lexical and context features into account for lexical simplification to a restricted vocabulary scope. The main contributions of the paper lie on: 1) leveraging WordNet-based semantic similarity measures to generate substitutions from community vocabulary; 2) utilizing Part-of-Speech tagging, stemming and word prototype matching to select substitutions to improve the quality of candidate words; 3) proposing a concise strategy for combining both n-gram and word2vec for semantic and context ranking to select the best candidates. Our experiments are based on a publicly available dataset containing 295 manually annotated sentences. The results show that our approach outperforms baseline methods, thus demonstrating its effectiveness in community-oriented lexical simplification tasks.

2   Related Work

Lexical simplification involves identifying complex words and seeking the best candidate substitutions for target words. The best substitution for a target word in a given sentence needs to be simpler while preserving the sentence grammatically and retaining its meaning as much as possible [11]. Typically, identification of the complex words incorporates: locating synonyms by various similarity measures; ranking and selecting the best suitable words based on criteria such as language model; and preserving the grammar and syntax of a sentence correctly [12].

There are some sub-tasks on lexical simplification. One of the widely known tasks is SemEval-2007 Task 10 [13]. The task involves both finding the synonyms and performing context disambiguation. In the task, an alternative substitute word or phrase for a target word in the text is identified. The data was selected from the English Internet Corpus of English without POS. Annotators are instructed to provide up to three equally good substitutes. They can also use slightly more general single words or phrases if their meanings are close. Systems mostly consider substitution candidates from WordNet and cast lexical substitution into a ranking task [14].

Many researches have been conducted on this task. Horn et al. [10] extracted over 30 K candidate lexical simplifications by identifying aligned words in a sentence aligned corpus of English Wikipedia with Simple English Wikipedia and proposed an approach utilizing the SVM candidate ranking. Baeza-Yates et al. [15] presented a context-aware method for lexical simplification that used free language resources and web n-gram frequencies. Paetzold et al. [3] proposed an unsupervised approach relying on a corpus of subtitles and a new type of word embedding model that accounts for the ambiguity of words. Alarcon et al. [16] provided support to the task of Complex Word Identification and selection of a simpler substitute by using easy-to-read resources.

In terms of semantic representation and computation, Wolf et al. [17] extended a word2vec framework to capture meaning across languages by representing words in both languages within a common semantic vector space. Sugathadasa et al. [18] introduced a domain specific semantic similarity measure created by the synergistic union of word2vec, which is commonly used for semantic similarity calculation and proposed a claim that word lemmatization which removes inflected forms of words can improve the performance of a word embedding model.

However, community-oriented lexical simplification is relatively unexplored. It is limited regardless of the target audience, as it tends to focus on certain steps of the simplification process and disregard others, such as the automatic detection of the words that require simplification. Paetzold et al. [3] focused on lexical simplification for English using non-native English speakers using an ensemble approach which achieves great performance in identifying words that challenge non-native speakers of English. The joint approach employed resource-light neural language models to simplify words deemed complex with pipelined approaches. Kajiwara et al. [19] proposed a method for acquiring plain lexical paraphrase using a Japanese dictionary in order to achieve lexical simplification for children. However, the work has not been compared with commonly used lexical simplification methods. Hao et al. [9] proposed a semantic-context combination ranking strategy for English lexical simplification to a restricted vocabulary. The strategy elaborately integrated commonly used semantic similarity calculation methods and context-based ranking methods as well as two POS-based matching methods, which were compared with commonly used lexical simplification methods and found to perform better on this problem. Their work can be referred to as the baseline for community-oriented lexical simplification in our experiment. However, their work needs further improvements: (1) Due to the limits of WordNet, candidate words of the same similarity were put randomly in the candidate list during the generation of the initial candidate words. Then the positions of the candidate words may result in ignoring the best words in the community list. (2) POS tagging used for the initial target words may filter out some of the best candidate words. (3) The candidate word may have the same stem as the target word, invalidating the replacement. (4) The context relevance calculating process for n-gram string extraction is inefficient due to the dataset used. The low coverage of grams combined with candidate words in the provided n-gram corpus leads to poor performance.

3   The Method

In the lexical simplification task, the substitution should retain both semantic meaning and grammatical structure of the original sentence being simplified. A new context-aware method based on hybrid ranking is proposed for community-oriented lexical simplification. The method generally consists of three steps. The first step is substitution candidate generation, which generates a list of candidate words, as $c_1, c_2, \ldots, c_n$, for the target word w without considering context of word. The second step is substitution selection, which selects the best candidates of the target word. The final step is substitution ranking, which re-ranks the selected candidates in terms of their simplicity. The overall framework of the proposed method is shown as Fig. 1.

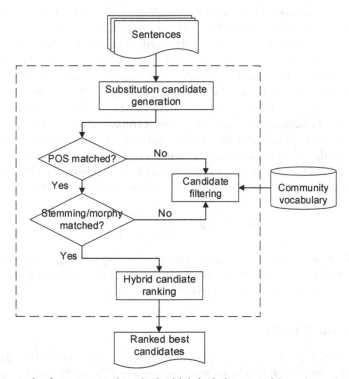

Fig. 1. Framework of our proposed method which includes part-of-Speech tagging, stemming, morphy matching, and semantic-context ranking for community-oriented lexical simplification.

3.1   Substitution Candidate Generation

Firstly, we determine the selection scope of initial candidate words since our task is lexical simplification for a specific community. Thus, the initial candidate words are obtained directly from a vocabulary list that related to the requirement of communities. The list of candidate words is then semantically expanded based on the initial candidate words using a synonym network. In order to improve the efficiency of identifying synonyms and keeping the grammatical appropriateness of target words and candidates, we apply part-of-speech (POS) tagging to target words and initial candidates for POS matching. After that, a semantic network that collects verbs, nouns, adjectives, and adverbs into synonym sets, such as WordNet, is applied to enlarge and link the list of substitutions based on word-based semantic relationships.

A semantic similarity measure based on WordNet is applied for the semantic relevance calculation between candidate words and target word. For a pair of words c_1 and c_2, the measure calculates the shortest path between c_1 and c_2. The calculation is based on the synonym set of target word and candidates, where each pair of words in these two synonym sets is calculated. The maximum similarity value is thus obtained after traversing the entire synsets and is used as the similarity between the target word and the candidate. The calculations using the path strategy is shown as Eq. (1), where $len(c_1, c_2)$ is the distance of the shortest path linking the two synsets. After the calculation, a similarity threshold is set to filter out candidate words with low similarity values. In the similarity calculation using synonym set, the similarities of some candidate words may be equal, and the randomly selected candidates can thus affect ranking performance. In order to further distinguish these candidate words, the Word2vec method is applied to calculate their relevance and then adjust their relative positions in generated candidate list from WordNet calculation. The Word2Vec model represents words into vectors based on several features such as vector dimensions and windows size. Therefore, Word2vec captures the similarity among words from the training of a commonly used large corpus, e.g., Google News. We preserve the candidate words and update their similarity values calculated by the Word2vec method and use the values for hybrid ranking.

$$Sim_{Path}(c_1, c_2) = \frac{1}{1 + len(c_1, c_2)} \tag{1}$$

After calculating context relevance, an answer-similarity set $A_{sim} = \{< a_m, a_{m, sim}>\}$ ($m = 1...n$) is sent to the rank aggregation process for further ranking, where $a_{m, sim}$ is question-answer cosine similarities of answer passage a_m.

3.2 Substitution Selection

To obtain fit substitutes for complex words, we propose a method which utilizes a set of filtering strategies together. The method applies word Synset Part-of-Speech (SYN-POS) matching and Target Word Part-of-Speech (TWPOS) matching before similarity calculation since POS labeling and matching can reduce the grammatical mismatching cases of candidate words and reduce the volume of candidates. In part-of-speech tagging, the difference in the part-of-speech expressions of adjectives has also an effect on the substitution performance. To improve the filtering effect of TWPOS, adjectives are transformed to adjective satellites in POS tagging of WordNet, e.g., a target word with the POS tag of ".a" is replaced by ".s", since satellite synset in an adjective cluster represents a concept that is similar in meaning to the concept represented by its head synset. For example, certain adjectives bind minimal meaning, e.g. "dry", "good", etc. Each of these is the center of an adjective synset in WordNet. Adjective satellites impose additional commitments on top of the meaning of the central adjective, e.g., "arid" equals to "dry" plus a particular context (i.e., climates). The POS transform process is shown as Eq. (2), where $pos(w_t)$ is the POS of the target word w_t, pos_{ADJ} represents POS tags with ".a" and pos_{ADJ_SAT} represents the POS tags with ".s".

$$pos(w_t) = \begin{cases} pos_{ADJ\_SAT} & pos(w_t) = pos_{ADJ} \\ pos(w_t) & pos(w_t) \neq pos_{ADJ} \end{cases} \tag{2}$$

In the generation process of candidate words, some generated words which are exactly the same as the target words and some words which have the same stem as the target word are kept in candidate lists. These candidate words are invalid as substitutions since they are simply different in form or in tense from the target word. Words with different singular and plural forms are of low frequency of occurrence in gold annotations since the complexity of the word structure is the same. For further filtering of the words generated by WordNet, different combination of lemmatization and stemming methods are explored. It was found that Porter stemming algorithm worked more effectively in extracting the stem information while morphy of WordNet method performed well in lemmatization. The filtering process is thus shown as Eq. (3), where w_t and w_c are the target word and candidate word, respectively, $stem(w_t)$ denotes the stemming of the word w_t by porter stemming tools and $lemma(w_t)$ denotes the lemmatization of the word w_t by the morphy method.

$$filter(w_c) = \begin{cases} 1 & stem(w_t) = stem(w_c) \text{ or } lemma(w_t) = lemma(w_c) \\ 0 & stem(w_t) \neq stem(w_c) \text{ and } lemma(w_t) \neq lemma(w_c) \end{cases} \tag{3}$$

3.3 Substitution Ranking

Based on the consideration that lexical substitution needs to ensure similar semantic meaning and grammatically fit to the original sentence, we calculate the similarity of candidate words according to their semantic relevance and filter out those that are semantically different from target words. After that, to minimize the influence of substitutions on the original meaning of a given sentence, candidate words are sorted based on their contextual relevance, which indicate whether a sentence is reasonable and fluent after the substitution. We mainly focus on the effective combination of these two processes. Firstly, we extract candidate strings as n-grams, where n is set to 2. Then the target words in the strings are replaced with candidate words to calculate the probability of the words in the strings. During every turn of calculations, the combinations of 2-g strings are retrieved from a reference corpus to get their frequencies. For example, for the target word "endure" in the sentence "Developers are routinely asked to endure hardships of design extremes", the extracted strings are "asked to", "to endure" and "endure hardships". Consequently, a candidate word "suffer" can be used to replace "endure" in the string to calculate its corresponding context relevance value. Finally, the top 10 candidates of the ranking list are selected as the best substitutions in this combined calculation strategy. For a candidate word w_c and its target word w_t in the sentence sen, the rank of the candidate word is calculated by the Eq. (4), (5), and (6).

$$R1(w_c) = Word2Vec\_similarity(w_c, w_t) \tag{4}$$

$$R2(w_c) = \frac{Match_{tw}(w_c, w_t)}{(1 - \beta)Sem_{syn}(w_c, w_t) + \beta\sqrt{Relv_{con}(w_c, sen)}} \tag{5}$$

$$Match_{tw}(w_c, w_t) = \begin{cases} 1 \ pos(w_c) = pos(w_t) \\ 0 \ pos(w_c) \neq pos(w_t) \end{cases} \tag{6}$$

In the equations above, the $Word2Vec\_similarity(w_c, w_t)$ is the similarity of w_c and w_t utilizing a group of related models that are used to produce word embeddings from word2vec. $Match_{tw}(w_c, w_t)$ is the binary value denoting the matching between the POS tags of w_c and w_t. $Sem(w_c, w_t)$ is the semantic similarity between the candidate word and the target word, while Sem_{syn} is the similarity with SYNPOS after the filtering with ξ. $Relv_{con}(w_c, sen)$ is the context fitting of the word w_c to the given sentence sen. The $Relv_{con}(w_c, sen)$ is defined as the maximum value of the frequency of a candidate word with surrounding context divided by its maximum frequency value. Here, the square root is applied to normalize the value range since the relevance value is usually small due to the large of maximum frequency. Additionally, we introduce a parameter β to balance the weights of similarity and context relevance in the final calculation.

4 Evaluation and Results

4.1 Data

We train the proposed method with a standard dataset [13] from SemEval 2007, which involves a lexical samples of nouns, verbs, adjectives and adverbs. As shown in Table 1, there are 295 sentences in total, extracted from the English Internet Corpus [20], and annotated by five native English speakers. Words in this lexical sample are selected to ensure variety of senses. Simultaneously, each senses of target word with the same part-of-speech has multiple instances. Finally, we evaluate the effectiveness of our method on Wikipedia dataset which contains 500 manually annotated sentences. The target word for every sentence is annotated by 50 independent annotators. We keep 249 sentences whose target words are not in the EDB list and whose gold answers are in the list, and name it as Dataset A. Then we further remove sentences whose annotation agreements of gold substitution candidates below 20%, i.e., at least 10 agreements of the gold substitutions of target word from the 50 independent annotators, and generate the Dataset B.

Table 1. The statistics of the datasets

Dataset	Counts	Average length per sentence	# candidates	Note
SemEval dataset	295 sentences	28.5 words	Top 10 of list	Target word with at least one instance
Wikipedia dataset	249 sentences	26.5 words	Top 10 of list	Target word with only one instance

4.2 Evaluation Metrics

We apply widely used metrics Accuracy @N [13], Best, and Oot (out-of-ten), as standard evaluation metrics in SemEval 2007 task.

Accuracy@N Measure. The metric determines whether any of the generated top N (N = 1, 2,... 10) candidate words are included in the gold set. If it is included, the current candidate word is marked as correct for a target word. The final accuracy is thus calculated as the number of correct matches divided by the total number of sentences. We calculate top 1 to top 10 accuracy scores, which show how often a gold-standard word is selected as the best fitting or among the 10 highest-ranked candidates and change of gold set coverage in this process.

Best Measure. The best candidates produced by the system which gives all words that the system believes are fitting, thus the credit for each correct guess is divided by the number of guesses. The first guess in the list with the highest overall rating is taken as the best guess. Best measures evaluate the quality of the best guess. We calculate recall and precision as the average annotator response frequency of substitutes found by the system over all items in dataset A and B. The metrics is represented as Eq. (7).

$$Precision_{best} = \frac{\sum_{a_i:i \in A} \frac{\sum_{res \in a_i} freq_{res}}{\frac{|a_i|}{|H_i|}}}{|A|}$$

$$Recall_{best} = \frac{\sum_{a_i:i \in T} \frac{\sum_{res \in a_i} freq_{res}}{\frac{|a_i|}{|H_i|}}}{|T|} \quad (7)$$

Oot Measures. Oot measure allows system to make up to 10 guesses, the credit for each correct guess is not divided by the number of guesses. With 10 guesses the system is more likely to find the responses of the gold annotations. The metrics is represented as Eq. (8).

$$Precision_{oot} = \frac{\sum_{a_i:i \in A} \frac{\sum_{res \in a_i} freq_{res}}{|H_i|}}{|A|}$$

$$Recall_{oot} = \frac{\sum_{a_i:i \in T} \frac{\sum_{res \in a_i} freq_{res}}{|H_i|}}{|T|} \quad (8)$$

4.3 The Result

The similarity threshold ξ for reducing the number of candidates is firstly trained and optimized. The performance using different ξ values are shown in Table 2. From the result, the performance on all the measures are same when ξ increases from 0.5 to 0.9. The accuracy achieves the best when ξ equals to 0.4, while other evaluation metrics reach the best when ξ equals to 0.3. Therefore, the optimized value of the parameter ξ is set to 0.3 based on overall consideration.

Table 2. Performance by setting ξ from 0 to 0.9 with the interval as 0.1 on the training set.

ξ	Accuracy										Best			Oot		
	@1	@2	@3	@4	@5	@6	@7	@8	@9	@10	P	R	F1	P	R	F1
0	0.169	0.315	0.359	0.407	0.468	0.492	0.525	0.529	0.539	0.566	0.097	0.097	0.097	0.37	0.37	0.37
0.1	0.193	0.322	0.386	0.424	0.495	0.508	0.536	0.536	0.539	0.566	0.103	0.103	0.103	0.37	0.37	0.37
0.2	0.193	0.319	0.38	0.424	0.492	0.512	0.536	0.536	0.536	0.569	0.118	0.118	0.118	0.371	0.371	0.371
0.3	**0.217**	**0.353**	0.383	0.444	0.485	0.502	0.508	0.525	0.556	**0.569**	**0.142**	**0.142**	**0.142**	**0.381**	**0.381**	**0.381**
0.4	0.2	0.339	**0.407**	**0.498**	**0.539**	**0.539**	**0.563**	**0.563**	**0.566**	0.566	0.114	0.114	0.114	0.368	0.368	0.368
0.5	0.203	0.349	0.386	0.431	0.492	0.495	0.505	0.515	0.515	0.519	0.131	0.127	0.129	0.361	0.347	0.354
0.6	0.203	0.349	0.386	0.431	0.492	0.495	0.505	0.515	0.515	0.519	0.131	0.127	0.129	0.361	0.347	0.354
0.7	0.203	0.349	0.386	0.431	0.492	0.495	0.505	0.515	0.515	0.519	0.131	0.127	0.129	0.361	0.347	0.354
0.8	0.203	0.349	0.386	0.431	0.492	0.495	0.505	0.515	0.515	0.519	0.131	0.127	0.129	0.361	0.347	0.354
0.9	0.203	0.349	0.386	0.431	0.492	0.495	0.505	0.515	0.515	0.519	0.131	0.127	0.129	0.361	0.347	0.354

After that, the semantic-context correlation parameter β for ranking is trained on training dataset. From the result, as shown in Fig. 2, the overall highest performance is achieved when ξ equals to 0.3 and β equals to 0.5. We thus use these values as the optimized parameters.

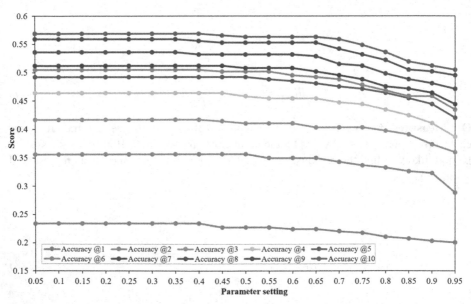

Fig. 2. Performance with different β values from 0.05 to 0.95 with the interval as 0.05 when ξ equals to 0.3.

To test the effects of gram length as the third parameter, we try different lengths of gram to compare their performance difference on the training dataset. A number of strategies using various lengths of grams from Google 1T n-gram corpus are evaluated using Accuracy@N. From the results, as shown in Table 3, it is most effective when the gram length is set to 2 which includes one word on both side of the being replaced word. It shows that the 2-g is more effective than 3-g. This may be caused by the larger matching coverage of the 2-g. We thus intentionally improve the coverage by judging whether two sets of 2-g containing the target word in 2-g library at the same time not in 3-g library. Eventually, the best combination of parameters is obtained as ($\xi = 0.3$, $\beta = 0.5$, $n = 2$).

Table 3. Performance with different gram lengths measured by Accuracy@1, by Accuracy@3, and by Accuracy@5 on the training set.

Strategies	Accuracy@1	Accuracy@3	Accuracy@5
Path	0.217	0.383	0.485
Path+Grank(2-grams)	0.227	0.410	0.492
Path+Grank(3-grams)	0.220	0.400	0.488

Table 4. Performance of our method and comparison with baseline methods on the Dataset A and Dataset B using Accuracy@10.

Methods	Accuracy on Dataset A									
	@1	@2	@3	@4	@5	@6	@7	@8	@9	@10
Four-step method	0.217	0.265	0.321	0.333	0.357	0.369	0.373	0.382	0.382	0.382
Word2Vector	0.120	0.133	0.137	0.137	0.137	0.137	0.137	0.137	0.137	0.137
Path+TWPOS+SYNPOS+Grank	0.201	0.289	0.329	0.357	0.378	0.398	0.410	0.410	0.414	0.426
Path+TWPOS+SYNPOS+Word2Vec+Grank	0.261	0.325	0.365	0.394	0.410	0.418	0.438	0.438	0.446	0.450

Methods	Accuracy on Dataset B									
	@1	@2	@3	@4	@5	@6	@7	@8	@9	@10
Four-step method	0.218	0.286	0.311	0.345	0.353	0.37	0.37	0.37	0.37	0.378
Word2Vector	0.176	0.176	0.176	0.176	0.176	0.176	0.176	0.176	0.176	0.176
Path+TWPOS+SYNPOS+Grank	0.218	0.311	0.319	0.361	0.403	0.420	0.420	0.420	0.420	0.420
Path+TWPOS+SYNPOS+Word2Vec+Grank	0.269	0.328	0.345	0.378	0.378	0.387	0.387	0.387	0.395	0.403

Using the combinations of the optimal parameters, we evaluate the performance of our lexical simplification model on the testing datasets, i.e., the Dataset A and Dataset B. The result of our method and the baseline methods using Accuracy@N are showed in Table 4, while the result using Best and Oot measures are showed in Fig. 3. In the comparison, our ranking method as Path+TWPOS+SYNPOS+Word2vec+Grank is compared with a list of baseline methods including the Four-step method, Word2vec, Path+TWPOS+SYNPOS+Grank. Four-step method uses WordNet synonyms based on four criteria in order, as the same baseline used in [13]. Word2vec is a two-layer neural net to group the vectors of similar words together in vector space for calculating similarity mathematically.

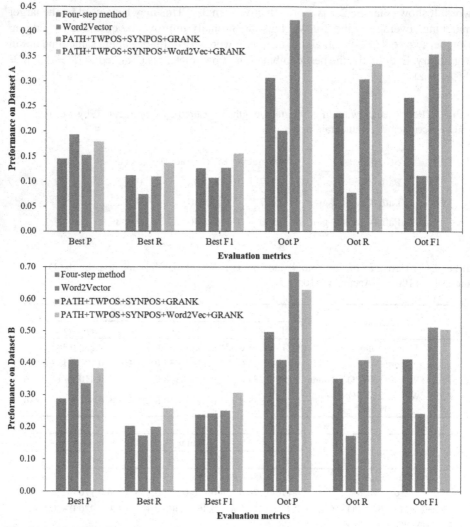

Fig. 3. Overall performance of our semantic-context combination ranking strategy and the baseline methods for community-oriented lexical simplification on Best and Oot metrics.

From the results as shown in Table 4, our approach achieves the best performance on all the Accuracy@N and Oot measures. Compared with the PathTW-POS+SYNPOS+Grank method, on the Dataset A, the performance of our method has improvement with Accuracy @1 from 0.201 to 0.261 (29.9%), Best P from 0.152 to 0.179 (17.8%) and Best F1 from 0.127 to 0.155 (22. 0%), while Oot P from 0.423 to 0.439 (3.8%) and Oot F1 from 0.354 to 0.380 (7.3%). On the Dataset B, our approach improved Accuracy @1 from 0.218 to 0.269 (23.4%), Best precision from 0.335 to 0.382 (14.0%), Best Recall from 0.2 to 0.257 (28.5%), best F1 from 0.25 to 0.307(22.8%). The reason for the slight decrease in accuracy @5-@10 and Oot are due to reordering of candidates, putting simpler candidate words for specified communities towards the front of the candidate list. At the same time, in the process, some candidate words that some annotators think are gold are ranked behind ten, resulting in a slight decrease in the result of Oot P. Overall, our method is still effective. Word2vec has highest Best P as 0.193 on Dataset A and 0.41 on Dataset B. However, the performances of using Word2vec on Best R and Oot measure are low, causing the low overall F1 scores. The results also demonstrate that the combination ranking approach achieves much better performance than any of the individual methods. It can be seen from the final comparison results that our model can achieve better lexical substitution for specific communities.

5 Conclusions

This paper proposes a context-aware lexical simplification approach based on a new ranking strategy combined with Word2vec and n-gram, which ranks according to the fitness in specific contexts around target words and similarities between target words and substitute candidates using a restricted vocabulary. A standard dataset is used for comparing our method with a list of baseline methods. Experiment results show that our proposed method achieves the best performance on substitute candidate generation for lexical substitution.

Acknowledgements. This work was supported by National Natural Science Foundation of China (No. 61772146), Natural Science Foundation of Guangdong Province (2018A030310051), and the Katie Shu Sui Pui Charitable Trust—Research and Publication Fund (KS 2018/2.8).

References

1. Saggion, H.: Automatic text simplification. In: Synthesis Lectures on Human Language Technologies, vol. 10(1). California, Morgan & Claypool Publishers (2017)
2. Kriz, R., Miltsakaki, E., Apidianaki, M.: Simplification using paraphrases and context-based lexical substitution. In: Conference of the North American Chapter of the Association for Computational Linguistics: Human Language Technologies (2018)
3. Paetzold, G.H., Specia, L.: Unsupervised lexical simplification for non-native speakers. In: Thirtieth AAAI Conference on Artificial Intelligence (2016)
4. Dagan, I., Glickman, O., Gliozzo, A., Marmorshtein, E., Strapparava, C.: Direct word sense matching for lexical substitution. In: Proceedings of the 21st International Conference on Computational Linguistics and 44th Annual Meeting of the Association for Computational Linguistics, pp. 449–456 (2006)
5. Dinu, G., Lapata, M.: Measuring distributional similarity in context. In: Proceedings of the 2010 Conference on Empirical Methods in Natural Language Processing, pp. 1162–1172. Association for Computational Linguistics (2010)
6. Thater, S., Fürstenau, H., Pinkal, M.: Contextualizing semantic representations using syntactically enriched vector models. In: Proceedings of the 48th Annual Meeting of the Association for Computational Linguistics, pp. 948–957. Association for Computational Linguistics (2010)
7. Melamud, O., Goldberger, J., Dagan, I.: Context2vec: learning generic context embedding with bidirectional LSTM. In: Proceedings of the 20th SIGNLL Conference on Computational Natural Language Learning (2016)

8. Education Bureau: Enhancing English Vocabulary Learning and Teaching at Secondary Level (2012). http://www.edb.gov.hk/vocab_learning_sec

9. Hao, T., Xie, W., Lee, J.: A semantic-context ranking approach for community-oriented english lexical simplification. In: Huang, X., Jiang, J., Zhao, D., Feng, Y., Hong, Y. (eds.) NLPCC 2017. LNCS (LNAI), vol. 10619, pp. 784–796. Springer, Cham (2018). https://doi.org/10.1007/978-3-319-73618-1_68

10. Horn, C., Manduca, C., Kauchak, D.: Learning a lexical simplifier using Wikipedia. In: Proceedings of the 52nd Annual Meeting of the Association for Computational Linguistics (Volume 2: Short Papers), pp. 458–463 (2014)

11. Qiang, J., Li, Y., Zhu, Y., Yuan, Y.: A Simple BERT-based approach for lexical simplification. arXiv preprint arXiv:1907.06226 (2019)

12. Swain, D., Tambe, M., Ballal, P., Dolase, V., Agrawal, K., Rajmane, Y.: Lexical text simplification using WordNet. In: Singh, M., Gupta, P.K., Tyagi, V., Flusser, J., Ören, T., Kashyap, R. (eds.) ICACDS 2019. CCIS, vol. 1046, pp. 114–122. Springer, Singapore (2019). https://doi.org/10.1007/978-981-13-9942-8_11

13. McCarthy, D., Navigli, R.: Semeval-2007 task 10: English lexical substitution task. In: Proceedings of the 4th International Workshop on Semantic Evaluations, pp. 48–53. Association for Computational Linguistics (2007)

14. Hintz, G., Biemann, C.: Language transfer learning for supervised lexical substitution. In: Proceedings of the 54th Annual Meeting of the Association for Computational Linguistics (Volume 1: Long Papers), pp. 118–129 (2016)

15. Baeza-Yates, R., Rello, L., Dembowski, J.: CASSA: a context-aware synonym simplification algorithm. In: Proceedings of the 2015 Conference of the North American Chapter of the Association for Computational Linguistics: Human Language Technologies, pp. 1380–1385 (2015)

16. Alarcon, R., Moreno, L., Segura-Bedmar, I., Martínez, P.: Lexical simplification approach using easy-to-read resources. Procesamiento del Lenguaje Natural **63**, 95–102 (2019)

17. Wolf, L., Hanani, Y., Bar, K., Dershowitz, N.: Joint word2vec networks for bilingual semantic representations. Int. J. Comput. Linguist. Appl. **5**(1), 27–42 (2014)

18. Sugathadasa, K., et al.: Synergistic union of word2vec and lexicon for domain specific semantic similarity. In: 2017 IEEE International Conference on Industrial and Information Systems (ICIIS), pp. 1–6. IEEE (2017)

19. Kajiwara, T., Matsumoto, H., Yamamoto, K.: Selecting proper lexical paraphrase for children. In: Proceedings of the 25th Conference on Computational Linguistics and Speech Processing, pp. 59–73 (2013)

20. Sharoff, S.: Open-source corpora: using the net to fish for linguistic data. Int. J. Corpus Linguist. **11**(4), 435–462 (2006)

Two-Level Convolutional Neural Network for Aspect Extraction

Jialin Wu[1], Yi Cai[1(✉)], Qingbao Huang[1,3], Jingyun Xu[1],
Raymond Chi-Wing Wong[2], and Jian Chen[1]

[1] School of Software Engineering, South China University of Technology,
Guangzhou, China
jlwu_scut@hotmail.com, {ycai,ellachen}@scut.edu.cn,
qbhuang@gxu.edu.cn, jingyun.x@qq.com
[2] Department of Computer Science and Engineering, The Hong Kong
University of Science and Technology, Hong Kong, China
raywong@cse.ust.hk
[3] School of Electrical Engineering, Guangxi University, Nanning, Guangxi, China

Abstract. Extract product aspect is an important task in fine-grained
sentiment analysis. Though many models have been proposed to solve the
problem. They either use handcrafted features or complex neural network
architectures. In this paper, we propose a simple Two-level CNN model
to extract product aspects from customer reviews, namely Char- and
Word-level CNN. The Char-level CNN learns char-level representation of
each word (also named morphological information), while the Word-level
CNN captures features from the concatenation of char-level representations and word embeddings. Compared to previous neural architectures,
our model do not use any external resources like dependency parsing tree
or lexicons. To the best of our knowledge, this is the first time to couple
Char- and Word-level CNN for aspect extraction. We conduct comparison experiments on two product review datasets. Experimental results
demonstrate the effectiveness of our proposed model.

Keywords: Aspect extraction · Convolutional network ·
Morphological information · Sequence labeling

1 Introduction

In sentiment analysis, more and more attention have been focused on Aspect-
based Sentiment Analysis (ABSA) [5,13]. ABSA aims to analyse the sentiments
expressed on the specific aspects. Thus, aspect extraction is a key task in ABSA,
it aims to extract aspects on which customers express their opinions. In product
reviews, aspects are attributes or features of an entity. Take the review "I charge
it at night and skip taking the cord with me because of the good battery life".
as an example, aspects in the review are "cord" and "battery life".

Aspect extraction has been studied for years. And there are mainly three
kind of methods for solving the problem, namely rule-based [5,22,24,30,36],

© Springer Nature Switzerland AG 2020
Y. Nah et al. (Eds.): DASFAA 2020 Workshops, LNCS 12115, pp. 93–105, 2020.
https://doi.org/10.1007/978-3-030-59413-8_8

unsupervised [12,17,19,28] and supervised methods [3,6,11,14,18,23,27,31,32]. Recently, people obtain state-of-the-art performances by applying deep learning models in many Natural Language Processing (NLP) tasks [2,8,34], like text classification, question answering and text summarization. In aspect extraction, existing works either use manual features or neural architectures that are complex designed [11,23,31,32], which are time consuming during feature engineering and model training. To address the aforementioned problems, we propose a CNN-based model without handcrafted features and complex architectures. We aims to create a simple architecture based on CNN [8,10] and make the model learn feature representations itself. In order to depend less on external knowledge like domain knowledge and syntactic parsing tree, we propose to learning the Char-level representations of words by a char-level CNN. And then couple that with word embeddings to get the final representation of words.

In the Char-level CNN network, we train the network to learning char-level representation of words. Although word embeddings [29] has been proved to be useful in many NLP tasks, it is learned based on the co-occurrence of words, and could not learn the morphological information [7] of each word. However, We think the morphological information is helpful. For example, the learned char-level representation of words with same suffix or prefix (like -ing, -less, etc) would be close if project them to a low dimensional space (like T-SNE [16]).

The Word-level CNN network is used to capture high-level features step by step. The input to it is the concatenation of char-level representations and word embeddings of words. We use multiple convolutional kernels with different kernel size. Although Recurrent Neural Network (RNN) and its variants, like Long Short-term Memory Network [4] and Gated Recurrent Network, have shown effectiveness in sequence labeling. One drawback of these RNNs is that each time step dependents on its previous or next time step (if bi-direction is used). The forward and backward propagation both serially go through the whole sequence which make the training time increase. We use CNN for its parallelization. In general, pooling operation is applied after convolution process. But in the Word-level CNN of our architecture, we do not use any pooling in order to align well with the input. We will give more details in the model section.

2   Related Work

Sentiment analysis (also called Opinion Mining) has been studied at three levels, namely document, sentence and aspect level [1,13,20]. We focus on aspect level for the other two levels are inaccuracy, while there are multiple aspects in a document or sentence. For instance, in the review "The service was good, but the price was expensive.", sentiment toward the aspect "service" is positive while "price" is negative. Therefore, inaccuracy results would be obtained if document or sentence level sentiment analysis is used, for they only give a total polarity toward the whole text. Thus, aspect level sentiment analysis is a more reasonable way as people may express different opinion toward different aspects of a product. One of the key tasks of aspect level sentiment analysis is aspect extraction. It aims to extract aspects on which customers express their opinions. It has been studied by many approaches. We classify these methods into three categories, namely rule-based, unsupervised and supervised approaches respectively.

Rule-based approach apply association rules, part-of-speech or syntactic dependency parsing tree results [24,30,36] to extract aspects. They thought aspects in consumer reviews are usually noun or noun phrases. [24] proposed a double propagation (DP) method to extract aspects from consumer reviews. It is based on the observation that there are natural relations between aspects and opinion terms due to the fact that opinion terms are used to modify aspects. Furthermore, they found that opinion terms and aspects themselves have relations in opinionated expressions too. Thus, they used a dependency parser based on the dependency grammar to identify these relations. As rules must be designed or chosen manually, it may not cover all the situations and is also time consuming. Besides, using external parsing tools may introduces additional errors.

The unsupervised approach includes methods such as topic modeling [12,17, 28] and label propagation [35]. The LDA model defines a Dirichlet probabilistic generative process for document-topic distribution. In each document, a latent aspect is chosen according to a multinomial distribution. [17] proposed a pLSA-based joint model, including topic model, positive sentiment model and negative sentiment model. [28] coupled features with structural information into topic model. However, topic modeling often only gives some rough topics in a document corpus rather than precise aspects, as a topical term dose not necessarily mean an aspect. For example, a topic model may find topical terms such as "battery", "life", "day" and "time" in a battery topic, these words are related to battery life but each individual word is not an aspect.

Supervised approach formulates the aspect extraction problem as a sequence tagging task. The common model including Conditional Random Fields (CRF) [9] and Hidden Markov Model (HMM) [25]. Feature engineering must be done carefully in these two models. Recently, deep learning models have been applied to many NLP tasks and achieved state-of-the-art performances. For aspect extraction, [32] proposed recursive neural conditional random fields to tagging words. [15] used Bi-directional LSTM-CNNs-CRF for end-to-end sequence labeling. Some multi-tasks learning models [11,31,32] were also proposed to co-extract aspects and opinion terms, using gold-standard opinion terms or sentiment lexicon.

Though the previous works could perform well in aspect extraction task, they either utilized external resources or designed complex neural architectures. In our work, we aim to propose a model without using external resources or domain knowledge, meanwhile, with simple network architecture. To address the first consideration, we propose to learn the char-level representations of words rather than using domain-dependent knowledge. To address the second consideration, we use a pure Convolutional Neural Network which is sequentially independent. The reason for this is that we want to make the model less dependent and as simple as possible, as in real-life application a complex model will slow down the speed of inference.

3 Model

3.1 Overview

We first describe the aspect extraction as a sequence labeling task. As an aspect can be a word or multi-word phrase, the task can be turned to identify the start and stop word of an aspect. This characteristic allows aspect extraction to be modeled as a sequence labeling task. Assume we have an input sentence of words $W = (\mathbf{w}_1, \mathbf{w}_2, \cdots, \mathbf{w}_m)$, where m is the number of words in the sentence. The task is to assign a label to each word $\mathbf{w}_i$. The label set is $\{B, I, O\}$, note that B indicates the beginning word and I indicates non-beginning word of an aspect phrase, while O indicates non-aspect words. Table 1 shows an example.

Table 1. An example of sequence labeling.

I	charge	it	at	night	and	skip	taking	the	cord	with	me	because	of	the	good	battery	life	.
O	O	O	O	O	O	O	O	O	B	O	O	O	O	O	O	B	I	O

The model we propose in this paper is a Two-level CNN (TCNN), including Char- and Word-level CNN. The Char-level CNN is for learning Char-level representation for words, while the Word-level CNN captures features between continuous words from the combination of their Char-level representation and word embeddings. Though word embeddings have shown great effectiveness in many NLP tasks, morphological information are not learned for it only considers co-occurrence of words. However, morphological information is indeed useful for identifying words with same prefix or suffix. And we think this feature can help to pose a valuable source of information for the task. Figure 1 and Fig. 2 show the overview of our model.

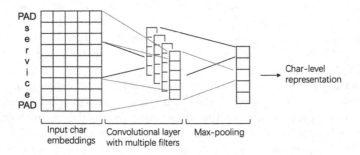

Fig. 1. Char-level CNN.

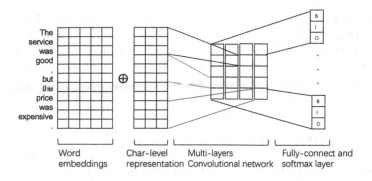

Fig. 2. Char- and Word-level CNN (TCNN).

3.2 Word-Level CNN

The proposed baseline model is a multi-layers convolutional neural network. It receives a word sequence of a sentence as input and predicts an output sequence of BIO tags. Note that we do not get rid of punctuations for they are helpful in identifying the boundary of an aspect.

First, the word sequence is passed to a word embedding layer to get the corresponding d^{wrd}-dimensional embedding vector $\mathbf{x}_i^{wrd}$ for the word $\mathbf{w}_i$ via an embedding matrix $\mathbf{W}^{wrd} \in \mathbb{R}^{d^{wrd} \times |V^{wrd}|}$:

$$\mathbf{x}_i^{wrd} = \mathbf{W}^{wrd} \mathbf{e}^{\mathbf{x}_i} \tag{1}$$

where V^{wrd} is the vocabulary of the word embeddings and $\mathbf{e}^{\mathbf{x}_i}$ is a one-hot vector of size $|V^{wrd}|$ representing the word $\mathbf{w}_i$. $\mathbf{W}^{wrd}$ could be randomly initialized or any pre-trained embeddings, like GloVe[1], Word2Vec[2] or Senna embedding[3].

The sequence of word embedding vectors is then passed to a 4-layers convolutional neural network. There are many 1-dimension filters with different kernel size $k = 2c + 1$ (k is an odd number) in each layer. The computation of a single convolutional layer is as follows:

$$\mathbf{o}_i = g\left(\sum_{z=-c}^{c} w_z \mathbf{x}_{i+z}^{wrd} + b \right) \tag{2}$$

where w_z indicates the weighting parameters of filters, b indicates bias and g is an activation function ReLU. $\mathbf{o}_i$ is the output of the layer. We pad the input sequence to the max sentence length and apply each filter to all the positions of the sequence. Note that we pad the left c and right c positions with all zeros so that the output of each layer is well-aligned with original input. In this way, we compute the representation for a word along with $2c$ nearby words in its context.

---

[1] https://nlp.stanford.edu/projects/glove/.
[2] http://code.google.com/archive/p/word2vec/.
[3] https://ronan.collobert.com/senna/.

We use two different kernel sizes in the first convolutional layer. For the rest 3 layers, we only employ one kernel size. At last, a Fully-connect and Softmax layer is applied across all positions to compute a probability distribution $\mathbf{q}_i$ over all possible output tags, namely **I**, **O** and **B** for each word. See the following formula:

$$\mathbf{q}_i = softmax(w^{tag}\mathbf{o}_i + b^{tag}) \tag{3}$$

where $\mathbf{o}_i$ is the feature vector for word $\mathbf{w}_i$. w^{tag} and b^{tag} are connection weighting and bias respectively. For each word, we select the tag with the highest probability as the predicted **IOB** tag. We add dropout layer after embedding layer and convolutional layers to avoid overfitting. Note that we do not use max-pooling in order to reserve the representation for each position and max-pooling would mix features across different positions.

3.3 Two-Level CNN (TCNN)

We propose a variation of the baseline model, which incorporates Char-level CNN. It is used to learn the Char-level representation of words, also known as morphological information. We treat a word as a sequence of chars. Assume given a char sequence $C = \{\mathbf{c}_1, \mathbf{c}_2, \cdots, \mathbf{c}_n\}$ of a word $\mathbf{w}$, we first get the corresponding d^{chr}-dimensional char embedding $\mathbf{x}_i^{chr}$ for char $\mathbf{c}_i$ via a char embedding matrix $\mathbf{W}^{chr} \in \mathbb{R}^{d^{chr} \times |V^{chr}|}$:

$$\mathbf{x}_i^{chr} = \mathbf{W}^{chr}\mathbf{e}^{c_i} \tag{4}$$

where V^{chr} is the char vocabulary. $\mathbf{e}^{c_i}$ is a one-hot vector of size $|V^{chr}|$ representing the char $\mathbf{c}_i$. The sequence of character embeddings is then passed through a convolutional layer, which consists of multiple filters with a fix kernel size k. After that, in order to get a fixed-size vector for the whole word, we apply max-pooling across each filter to select the most important feature. Finally, we can get a char-level representation $\mathbf{x}^c$ for the word.

To incorporate the baseline model in the overall neural network model, we first pass the corresponding char sequence of each word in $W = \{\mathbf{w}_1, \mathbf{w}_2, \cdots, \mathbf{w}_m\}$ to our Char-level CNN to get $X^c = \{\mathbf{x}_1^c, \mathbf{x}_2^c, \cdots, \mathbf{x}_m^c\}$. These Char-level representations are then concatenated with the word embeddings. See the following computation:

$$\mathbf{x}_i^{cwrd} = \mathbf{x}_i^{wrd} \bigoplus \mathbf{x}_i^c \tag{5}$$

$\bigoplus$ indicates concatenation of two vectors. And this is the new input of our baseline model. As $\mathbf{x}_i^{cwrd}$ contains Char- and Word-level information, it can help the subsequent CNN to learn more useful features for aspect extraction.

3.4 Network Training

In our model, the trainable parameters are w_z, b, $\mathbf{W}^{chr}$, w^{tag}, b^{tag}. We do not allow $\mathbf{W}^{wrd}$ trainable, for small training examples may lead to many unseen words in test data. If we set word embeddings trainable, the embeddings of unseen words in the test data still have the old features, while those seen words in the train data are adjusted and CNN filters will adjust to the new features accordingly. This can result in the old features are mistakenly extracted by CNN. The optimization of the model parameters is done by minimizing the classification error for each word in the sequence. The optimization is carried out using a mini-batch size of 64 with the stochastic optimization technique *Adam*. We fit the model by minimizing the negative log likelihood (NLL) of the training data. The NLL for the sentence **s** can be written as:

$$J(\theta) = \sum_{t=1}^{T} \sum_{k=1}^{K} y_{tk} log P(y_t = k|s, \theta) \tag{6}$$

where $y_{tk} = I(y_t = k)$ is an indicator variable to encode the gold labels, i.e., $y_{tk} = 1$ if the gold label $y_t = k$, otherwise 0. T denotes the total number of the words in the sentence and K is the length of labels set. The loss function minimizes the cross-entropy between the predicted distribution and the target distribution.

We use PyTorch to implement our model and train it on a Tesla K80 GPU server. It took about 25 min to run 200 epochs for one training data and about 5 s to infer testing data.

4   Experiments

In this section, we present the datasets and evaluation used for the task of aspect extraction from online customer reviews. Besides, baseline methods and experiment settings will be given to compare our model. Finally, we discuss our experiment results to show that the proposed TCNN is more effective compared to previous approaches, and the Char-level CNN plays an important role in our model.

4.1   Datasets and Evaluation

In our experiments, following the recent paper [11], we use two review datasets provided by SemEval-2014[4] and SemEval-2016[5] respectively. The first dataset is from laptop domain on subtask 1 of SemEval-2014 Task 4. The second dataset is from the restaurant domain on subtask 1 (slot 2) of SemEval-2016 Task 5. These datasets consist of review sentences with aspect terms labeled as spans of characters. Statistics of the two datasets are presented in Table 2.

We also investigate the number of words that make up an aspect term. And the basic statistics of the datasets are presented in Table 3. We can see that the majority of aspects have only one word, while about one third of them have

---

[4] http://alt.qcri.org/semeval2014/task4/.
[5] http://alt.qcri.org/semeval2016/task5/.

Table 2. Number of sentences (#.S.) and aspect (#.A.) in dataset

Dataset	Training #S./#A.	Testing #S./#A.
SemEval-14 Laptop	3045/2358	800/654
SemEval-16 Restaurant	2000/1880	676/650

Table 3. Number of one-word and multi-word aspect in dataset

	SemEval-14 Laptop	SemEval-16 Restaurant
	Training/Testing	Training/Testing
One-word aspects	1493/364	1373/485
Multi-word aspects	865/290	507/165
Total aspects	2358/654	1880/650

multiple words. Besides, some sentences in the datasets have more than one aspect while some have no aspect.

Commonly, the metric for aspect extraction is $F1$ score (also called F-Measure). It considers both the precision p and the recall r. as shown in Formula (6):

$$F1 = 2 \cdot \frac{p \cdot r}{p + r} \tag{7}$$

where p is the number of correct positive results divided by the number of all positive results returned by the model, and r is the number of correct positive results divided by the number of all samples that should have been identified as positive.

4.2   Settings

We hold out a small part of the training set as validation data to decide the hyper-parameters of our model.

In the Char-level CNN, the max char number per word are 20 and 17 for Laptop and Restaurant dataset respectively. The char embedding matrix $\mathbf{W}^{chr}$ is initialized by uniform distribution with range $[-\sqrt{\frac{3}{d^{chr}}}, \sqrt{\frac{3}{d^{chr}}}]$ and made trainable during the training procedure. d^{chr} is set to 100. The convolutional layer has 128 filters with kernel size $k = 3$. Max-pooling is then applied across each filter to get a 128 dimensions vector for each word.

In the Word-level CNN, the max length of sentence is 83. We use three kinds of pre-trained embedding for comparison. They are GoogleNews, GloVe42B and GloVe840B embeddings with 300 dimensions. The first convolutional layer has 128 filters with kernel size $k = 3$ and 128 filters with kernel size $k = 5$. The rest 3 convolutional layers has 256 filters with kernel size $k = 5$. We use Adam optimizer with learning rate 0.0001 to minimize the NLL loss function and the dropout rate is 0.5.

In order to prove the importance of the Char-level CNN, we conduct experiments which only use Word-level CNN. And compare it with TCNN to see whether TCNN can obtain better results.

Generally, Conditional Random Field (CRF) is applied to sequence labeling task. In our work, we also add CRF to our TCNN model to see whether it can improve the results.

4.3   Baseline Methods

We compare our model with two groups of baselines using the datasets aforementioned.

The first group uses single-task methods:

CRF, with manual features and Glove word embedding [21].

WDEmb, proposed by Yin [33], coupled word embeddings, linear context embeddings and dependency path embeddings with CRF.

LSTM, used in [14] and [11].

BiLSTM-CNN-CRF, proposed by Reimers and Gurevych [26] for Name Entity Recognition.

The second group uses multi-task learning:

RNCRF [31], a joint model for aspect and opinion terms co-extraction using dependency tree based recursive neural network and CRF, with handcrafted features.

CMLA [32], a multi-layer coupled-attention network that also performs aspect and opinion terms co-extraction.

MIN [11], a multi-task learning framework with three LSTMs, two for co-extraction of aspects and opinions, while the last one for discriminating sentimental and non- sentimental sentences.

4.4   Results and Analysis

Table 4 and Table 5 show the results for Laptop and Restaurant datasets with different settings respectively. We can see that, compared to the model only uses Word-level CNN, TCNN improves the results obviously, which demonstrates the effectiveness of Char-level CNN. The later visualizations will also show that it can really learn morphological information from words. It is well known that CRF is suitable for sequence labeling, but as shown in tables, TCNN with CRF do not obtain better results for many aspects are just single word and CRF may be useful when aspects are multiple words. We can find that GloVe840B embeddings performs better than GoogleNews and GloVe42B embeddings, which demonstrates pre-trained embeddings from a larger corpus can performs better

Table 4. $F1$ for Laptop with different settings

	GoogleNews	GloVe-42B	GloVe-840B
Word-level CNN	70.97	66.06	79.06
TCNN	74.59	74.55	**80.46**
TCNN+CRF	74.02	72.49	79.17

Table 5. $F1$ for Restaurant with different settings

	GoogleNews	GloVe-42B	GloVe-840B
Word-level CNN	67.10	64.94	72.20
TCNN	69.52	68.42	**73.52**
TCNN+CRF	68.97	68.12	72.83

than that from a smaller one. Also, we can see that, as corpus grows larger, TCNN obtains less improvements compared to Word-level CNN, which proves the importance of pre-trained word embeddings.

Table 6 shows comparison results among different models. The results of the first two groups are copied from [11]. We can see that GloVe840B-TCNN performs the best compared to both single-task and multi-task methods. From the multi-task learning approaches we can observe that they obtain results higher than single-task methods, which demonstrates multi-task learning can really bring more information for each task. But our Two-level CNN with pre-trained embeddings can even get better results, which once again proves the usefulness of our proposed model.

Table 6. Comparison results in F_1 score

Model	Laptop	Restaurant
CRF	74.01	69.56
WDEmb	75.16	–
LSTM	75.25	71.26
BiLSTM-CNN-CRF	77.8	72.5
RNCRF	78.42	–
CMLA	77.80	–
MIN	77.58	73.44
Glove840B-TCNN	**80.46**	**73.52**

In order to confirm that the Char-level CNN can really learn morphological information for words, we collect outputs of the Char-level CNN for words with same suffix and use T-SNE to project them into a 2 dimensional space. The result is shown in Fig. 3. We can find that words with same suffix (like -ly, -less, -ing, -able) are grouped in the low dimensional space. For comparison, we also use the corresponding GloVe embedding of those words to see what can be learned. The result is shown in Fig. 4, we can see that there is no clear separation between the four suffix groups. It once again proves the effectiveness of Char-level CNN.

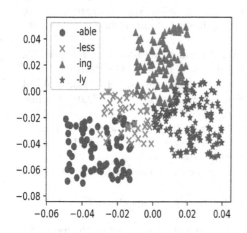

Fig. 3. Char-level embeddings.   **Fig. 4.** GloVe embeddings.

5   Conclusion

In this paper, we propose a Two-level CNN model for aspect extraction. Without using external resources like dependency parsing tree, lexicon and hand-crafted features, our model achieves competitive results compared to previous approaches. We conduct a series of comparative experiments with different settings, like different kinds of pre-trained word embeddings. We also add CRF to see whether it could improve the result. We find that Two-level CNN with GloVe840B embeddings performs best without using CRF. After that, we compare our model with two groups previous methods. Experimental results show the effectiveness of our model compared to single- and multi- task methods.

Acknowledgement. This work was supported by the Fundamental Research Funds for the Central Universities, SCUT (No. 2017ZD048, D2182480), the Science and Technology Planning Project of Guangdong Province (No. 2017B0- 50506004), the Science and Technology Programs of Guangzhou (No. 2017040300-76, 201802010027, 201902010046) and the Guangxi Natural Science Foundation (No. 2017GXNS-FAA198225).

References

1. Cambria, E., Hussain, A.: Sentic Computing: Techniques, Tools, and Applications, vol. 2. Springer Science & Business Media, Heidelberg (2012)
2. Gehring, J., Auli, M., Grangier, D., Yarats, D., Dauphin, Y.N.: Convolutional sequence to sequence learning (2017)
3. He, R., Lee, W.S., Ng, H.T., Dahlmeier, D.: An unsupervised neural attention model for aspect extraction. In: Proceedings of the 55th Annual Meeting of the Association for Computational Linguistics (Volume 1: Long Papers). vol. 1, pp. 388–397 (2017)
4. Hochreiter, S., Schmidhuber, J.: LSTM can solve hard long time lag problems. In: Advances in Neural Information Processing Systems, pp. 473–479 (1997)
5. Hu, M., Liu, B.: Mining and summarizing customer reviews. In: Tenth ACM SIGKDD International Conference on Knowledge Discovery and Data Mining, Seattle, Washington, USA, pp. 168–177, August 2004
6. Jakob, N., Gurevych, I.: Extracting opinion targets in a single- and cross-domain setting with conditional random fields. In: Conference on Empirical Methods in Natural Language Processing, pp. 1035–1045 (2010)
7. Jebbara, S., Cimiano, P.: Improving opinion-target extraction with character-level word embeddings (2017)
8. Kim, Y.: Convolutional neural networks for sentence classification. Eprint Arxiv (2014)
9. Lafferty, J., McCallum, A., Pereira, F.C.: Conditional random fields: Probabilistic models for segmenting and labeling sequence data (2001)
10. Lecun, Y., Bengio, Y.: Convolutional Networks for Images, Speech, and Time Series. MIT Press, Cambridge (1998)
11. Li, X., Lam, W.: Deep multi-task learning for aspect term extraction with memory interaction. In: Proceedings of the 2017 Conference on Empirical Methods in Natural Language Processing, pp. 2886–2892 (2017)
12. Lin, C., He, Y.: Joint sentiment/topic model for sentiment analysis. In: ACM Conference on Information & Knowledge Management, pp. 375–384 (2009)
13. Liu, B.: Sentiment analysis and opinion mining. Synth. Lect. Hum. Lang. Technol. **5**(1), 1–167 (2012)
14. Liu, P., Joty, S., Meng, H.: Fine-grained opinion mining with recurrent neural networks and word embeddings. In: Proceedings of the 2015 Conference on Empirical Methods in Natural Language Processing, pp. 1433–1443 (2015)
15. Ma, X., Hovy, E.: End-to-end sequence labeling via bi-directional lstm-cnns-crf. arXiv preprint arXiv:1603.01354 (2016)
16. Maaten, L.V.D., Hinton, G.: Visualizing data using t-SNE. J. Mach. Learn. Res. **9**, 2579–2605 (2008)
17. Mei, Q., Ling, X., Wondra, M., Su, H., Zhai, C.X.: Topic sentiment mixture: modeling facets and opinions in weblogs. In: International Conference on World Wide Web, pp. 171–180 (2007)
18. Mitchell, M., Aguilar, J., Wilson, T., Van Durme, B.: Open domain targeted sentiment. In: Proceedings of the 2013 Conference on Empirical Methods in Natural Language Processing, pp. 1643–1654 (2013)
19. Moghaddam, S., Ester, M.: ILDA: interdependent LDA model for learning latent aspects and their ratings from online product reviews. In: Proceedings of the 34th International ACM SIGIR Conference on Research and Development in Information Retrieval, pp. 665–674 (2011)

20. Pang, B., L, Lee., et al.: Opinion mining and sentiment analysis. Found. Trends®
    Inf. Retrieval **2**(1), 1–135 (2008)
21. Pennington, J., Socher, R., Manning, C.: Glove: global vectors for word represen-
    tation. In: Proceedings of the 2014 Conference on Empirical Methods in Natural
    Language Processing (EMNLP), pp. 1532–1543 (2014)
22. Popescu, A.M.: Extracting product features and opinions from reviews. In:
    HLT/EMNLP on Interactive Demonstrations, pp. 32–33 (2005)
23. Poria, S., Cambria, E., Gelbukh, A.: Aspect extraction for opinion mining with a
    deep convolutional neural network. Knowl. Based Syst. **108**, 42–49 (2016)
24. Qiu, G., Liu, B., Bu, J., Chen, C.: Opinion word expansion and target extraction
    through double propagation. Comput. Linguist. **37**(1), 9–27 (2011)
25. Rabiner, L.R.: A tutorial on hidden Markov models and selected applications in
    speech recognition. Proc. IEEE **77**(2), 257–286 (1989)
26. Reimers, N., Gurevych, I.: Reporting score distributions makes a difference: per-
    formance study of LSTM-networks for sequence tagging (2017)
27. Shu, L., Xu, H., Liu, B.: Lifelong learning crf for supervised aspect extraction
    (2017)
28. Titov, I., McDonald, R.: A joint model of text and aspect ratings for sentiment
    summarization. In: Proceedings of ACL-08: HLT, pp. 308–316 (2008)
29. Turian, J., Ratinov, L., Bengio, Y.: Word representations: a simple and general
    method for semi-supervised learning. In: Proceedings of the 48th Annual Meeting
    of the Association for Computational Linguistics, pp. 384–394. Association for
    Computational Linguistics (2010)
30. Wang, B., Wang, H.: Bootstrapping both product features and opinion words from
    Chinese customer reviews with cross-inducing. In: Proceedings of the Third Inter-
    national Joint Conference on Natural Language Processing: Volume-I (2008)
31. Wang, W., Pan, S.J., Dahlmeier, D., Xiao, X.: Recursive neural conditional random
    fields for aspect-based sentiment analysis (2016)
32. Wang, W., Pan, S.J., Dahlmeier, D., Xiao, X.: Coupled multi-layer attentions for
    co-extraction of aspect and opinion terms. In: AAAI, pp. 3316–3322 (2017)
33. Yin, Y., Wei, F., Dong, L., Xu, K., Zhang, M., Zhou, M.: Unsupervised word and
    dependency path embeddings for aspect term extraction, pp. 2979–2985 (2016)
34. Zhang, X., Zhao, J., Lecun, Y.: Character-level convolutional networks for text
    classification. In: Advances in Neural Information Processing Systems, pp. 649–
    657 (2015)
35. Zhou, X., Wan, X., Xiao, J.: Collective opinion target extraction in Chinese
    microblogs. In: Proceedings of the 2013 Conference on Empirical Methods in Nat-
    ural Language Processing, pp. 1840–1850 (2013)
36. Zhuang, L., Jing, F., Zhu, X.Y.: Movie review mining and summarization. In: Pro-
    ceedings of the 15th ACM International Conference on Information and Knowledge
    Management, pp. 43–50 (2006)

Leveraging Statistic and Semantic Features for Similar Question Detection Using Fusion XGBoost

Siyuan Liao[1], Leung-Pun Wong[2], Lap-Kei Lee[2], Oliver Au[2],
and Tianyong Hao[1(✉)]

[1] School of Computer Science, South China Normal University,
Guangzhou, China
{sean,haoty}@m.scnu.edu.cn
[2] School of Science and Technology, The Open University of Hong Kong,
Hong Kong, China
{s1243151,lklee,oau}@ouhk.edu.hk

Abstract. Question text similarity calculation is a fundamental and essential research problem for community question answering services. Different question text collections have various characteristics. Some frequently answered questions may have distinct statistical patterns, while some questions are syntactically different but semantically similar. To measure question similarity more adaptively to different kinds of question text, this paper proposes a method for identifying similar question utilizing the combination of both statistic and semantic features based on XGBoost. The method extracts semantic and statistical features from question text. After that, a feature set generation method is proposed, along with a model fusion strategy. Based on the standard Yahoo! dataset containing 25,569 questions with answers, three experiments have been conducted to evaluate the performance of the method. Results show that it achieves a precision of 88.65% and a recall of 71.85% outperforming a list of baseline methods.

Keywords: Similar question detection · Feature set generation · XGBoost · Question-answering

1 Introduction

In recent years, Question Answering (QA) has become an important research field of Natural Language Processing (NLP). Detecting similar question is a fundamental problem for Community Question Answering (CQA) systems such as Yahoo! Answers, Baidu Knows and Quora. These systems serve as a platform for users to share their knowledge, allowing users to submit questions or answer questions posted by other users. Due to their popularity, a lot of questions and answers have been recorded and accumulated. The possibility of reusing existing frequently asked questions (FAQ) makes QA systems more convenient and

© Springer Nature Switzerland AG 2020
Y. Nah et al. (Eds.): DASFAA 2020 Workshops, LNCS 12115, pp. 106–120, 2020.
https://doi.org/10.1007/978-3-030-59413-8_9

adaptable. If a question similar to a given new question is identified in the collection of FAQ, corresponding answers can be easily provided by retrieving existing answers associated with the question.

However, the major challenge associated with similar question retrieval is lexico-syntactic gap. Two questions may refer to the same thing but they may differ lexically and syntactically. Thus, many evaluation tasks have been proposed in academia, e.g., SemEval [14], or industries, e.g., Quora[1], where their solutions can become components of automatic methods for detecting duplicate questions.

At present, the main research methods are divided into two categories: traditional machine learning methods and deep learning methods. The performance of traditional methods mainly depend on text representation and similarity metrics, e.g. tf-idf [17] and Word2Vec [13]. Deep learning based models, e.g., [4,16,24], use variants of neural network architectures to model question-question pair similarity. Although deep learning based models have shown their strong ability and convenience in the task, it is hard to achieve the desired accuracy as large training sets are typically not available. Moreover, effectively exploiting full-syntactic parse information in Neural Networks is still an open problem. Traditional methods mainly rely on feature engineering. Based on previous experience, the use of a single similarity metric is prone to incorrect classification of some question pairs.

In this paper, we propose an automated method for similar question detection based on Fusion XGBoost. In order to solve the problem caused by a single feature model, a feature combination method is used to generate candidate feature sets from statistical features and semantic features. We use selected candidate feature sets as input to train the XGBoost model separately, and then adopt a multi-model fusion method based on voting mechanism to enhance the generalization performance of the model. The main contribution of this work lies in two aspects: (1) A statistical and semantic similarity feature combination method is proposed, by which the feature sets generated are more effective. (2) A new model fusion method via XGBoost is proposed, which utilized multiple feature sets.

2 Related Work

Different approaches have been proposed to identify similar questions in CQA. The traditional measures of similarity mainly used metrics based on word frequency calculation and part-of-speech (POS) tagging. Niwattanakul et al. [15] proposed a similarity measurement between keywords and index terms based on Jaccard Coefficient. Huang et al. [8] proposed measurement method of similarity based on tf-idf. Hao et al. [6] proposed an automated approach to detecting similar questions based on the calculation of question topical diversity. Zhou et al. [23] adopted a phrase-based translation model on Yahoo! Answers and evaluated its effectiveness. Wang et al. [21] proposed a model to find semantically

---

[1] https://www.kaggle.com/c/quora-question-pairs.

related questions by computing similarity between the representing syntactic trees of questions.

There are also a lot of research on measuring semantic similarity of texts. Jiang et al. [9] devised a semantic similarity metric using WordNet. Li et al. [11] devised another semantic similarity metric using HowNet. Li and Zhao [12] used the ontology of domain to calculate semantic similarity. Based on the concept of distributed word vectors, the Word2Vec technique proposed by Mikolov et al. [13] has been sucessfully applied in many NLP tasks. After that, pre-training with large-scale corpora and then tuning on specific tasks has become a mainstream of obtaining semantic representation. Kusner et al. [10] applied word mover's distance to discuss the similarity of two documents. Devlin et al. [5] proposed a new pre-training language representation method and obtain contextual embeddings for sentences, which have been applied to sentence similarity calculation.

Some existing work on detecting similar questions is based on deep learning. Zhou et al. [24] adopt a deep neural network (DNN) based query and answer representation model to rank a set of answers for a given query. Qiu and Huang [16] proposed a convolutional neural tensor network architecture to encode the sentences in semantic space and achieved semantic matching. Das et al. [4] proposed a deep structure topic model to bridge the lexico-syntactic gap between questions. Ruan et al. [18] discussed the methods of calculating the similarity of sentences based on multi-feature fusion. Ye et al. [22] used recurrent neural network (RNN) to measure the semantic similarity between sentences. Chali and Islam [2] applied Long Short-term memory (LSTM) and bi-direction Long Short-term memory (biLSTM) to find the semantic similarity between questions. Uva et al. [20] proposed to inject structural representations in Neural Networks for solving question similarity.

Each method mentioned above has its advantage, but they deal with only a few aspects of sentences. Obviously, questions well matched on different similarity measures are more likely to be similar. Based on this assumption, Song et al. [19] proposed to employ both statistic measure and semantic information. The similarity measure was calculated by combination of dynamically formed vectors and WordNet. Different from the method of simple feature weighting, we propose feature set generation and model fusion methods to utilize multiple different feature sets.

3   Our Approach

The framework of the proposed fusion XGBoost model mainly contains four parts: data preprocessing, feature extraction, feature set generation and model fusion. The overall framework is shown in Fig. 1.

In order to effectively obtain the statistic and semantic features of questions, several data preprocessing strategies are applied. First we remove useless characters other than letters and numbers, unify some common phrases into the same form, and then perform word segmentation and spell correction. After that, statistic and semantic similarity features are extracted from clean question text.

These two types of features are used to initialize the feature sets. The feature set generation algorithm traverses the initial feature sets to generate candidate feature sets. Then, the XGBoost model is trained by selecting feature sets from the candidate feature sets according to the rules of model fusion. After that, a voting fusion based on F1 score is adopted. Finally, the final similar question results are obtained using the fusion model prediction.

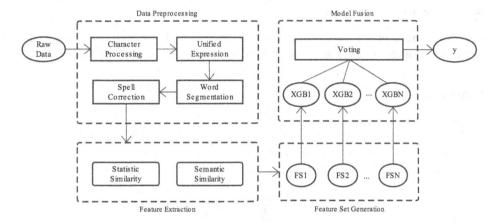

Fig. 1. Overall framework of fusion XGBoost

3.1 Data Preprocessing

For traditional features based on statistical word frequency, the biggest challenge is word sense disambiguation, that is, the same meaning may have different expressions, e.g., different tenses of verbs, singular or plural form of nouns, etc. These differences can be ignored when measuring the similarity of questions. We first remove all characters except letters, numbers and punctuation in the dataset, and then unify all the letters into lowercase.

Unified Expression. Certain phrases have different expressions, such as *"what's"* and *"what is"*. The two expressions with the same meaning do not need to be distinguished into different phrases when using word frequency based features. To this end, we adopt a unified expression approach, using regular expressions to unify phrases such as *"what's"*, *"can't"* and *"world cup"*, replaced with *"what is"*, *"cannot"*, *"worldcup"*. On the other hand, there are Internet catchwords and special domain names in daily life, such as *"4all"*, *"any1"*, *"6x"*, etc. The commonality is that they are all combinations of letters and numbers. We split them and got *"4 all"*, *"any 1"*, *"6 x"* as the result.

Spell Correction. Text data entered by users on web may have various spelling errors. For example, *"guitars"* is spelled *"guitar"*, *"perfect"* is spelled *"perrfect"*, etc. We adopt an open source library pyspellchecker[2] to correct the spelling issues. It applies a Levenshtein Distance algorithm [7] to find permutations within an edit distance of 2 from original word, and then compares all permutations (insertions, deletions, replacements, and transpositions) to known words in a word frequency list. Those words that are found more often in the frequency list are more likely to be the correct results.

3.2 Feature Extraction

The calculation of sentence similarity utilizes a list of similarity metrics. The metrics focusing on either statistic or semantic aspects are used as features.

I. Statistic Similarity Metrics

1) Jaccard Index (Jac)

   The Jaccard Index treats a sentence as a collection of words, and then divides the number of intersection elements between two sentences by the number of union elements to get the similarity between the two sentences:

$$Jac(A, B) = \frac{|A \cap B|}{|A \cup B|} = \frac{|A \cap B|}{|A| + |B| - |A \cap B|},$$ (1)

   where A, B are respectively the collections of words for the two questions without stop words.

2) Word's Common Rate (CR)

   Word's Common Rate is a kind of basic metric to describe the similarity between two collections:

$$CR(A, B) = \frac{|A \cap B|}{|A| + |B|}.$$ (2)

3) TF-IDF via Word Match (TWM)

   Let $Diff(A)$ denote the word in question A but not in question B, and $Diff(B)$ denote the word in B but not in A. $tfidf(a)$ denote the TF-IDF value of word a. The TWM is calculated using Eq. (3):

$$TWM(A, B) = \frac{\sum_{i=1}^{|C|} tfidf(C_i)}{\sum_{j=1}^{|D|} tfidf(D_i)},$$ (3)

   where $C = Diff(A) \cup Diff(B)$ and $D = A \cup B$.

4) Topical Diversity (TD)

   Topical words can be generated with different POS tags for a given question. Four combinations of POS tabs are defined as: POS1 (nouns), POS2 (nouns and adjectives), POS3 (nouns and verbs), POS4 (nouns, adjectives and verbs).

---

[2] https://github.com/barrust/pyspellchecker.

After generating topical words, topical diversity between two questions can be calculated by:

$$TD(A, B) = 1 - \frac{\sum_{t_i \in T_A \cap T_B} tfidf(t_i)}{\sum_{t_j \in T_A \cup T_B} tfidf(t_j)}, \tag{4}$$

where A and B are the two questions, T_A and T_B are the collections of topic words in question A and question B, respectively. From this equation, we can observe that the topical diversity is higher when there are less shared topical words between the questions.

II. Semantic Similarity Metrics

1) Cosine similarity via BERT Sentence Embedding (CosBERT)
   BERT pre-training model is adopted to extract the sentence embedding of a given question. The pre-training model is provided by Google Research [5], which gives vector to each word and performs pooling for sentence embedding. The cosine similarity is calculated as:

$$Cosine(V_A, V_B) = \frac{V_A * V_B}{\|V_A\|_2 * \|V_B\|_2} \tag{5}$$

   where V_A and V_B are vectors for word A and B, respectively. $\|\cdot\|$ is Euclidean norm.

2) Cosine similarity via Word2Vec (CosW2V)
   The pre-trained word2vec model gives word embedding $V(w)$ for each word in sentence. The sentence embedding for a sentence containing N words w_1, w_2, ..., w_N is calculated as:

$$V = \frac{\sum_{i=1}^{N} V(w_i)}{\sqrt{\sum_{i=1}^{N} V(w_i)^2}} \tag{6}$$

   After that, cosine similarity between two questions can be obtained through formula (5).

3) Cosine similarity via Smooth Inverse Frequency weighted Word2Vec (CosSIF)
   In order to extract semantic information, the word vector in a sentence is usually averaged as a sentence vector. Averaging gives too much weight to irrelevant words. Arora et al. [1] proposed an algorithm for embedding sentence using word embedding, which uses the smooth inverse frequency feature of the term to weight the word embedding and then average it. After generating all questions' sentence embeddings, PCA/SVD is applied to modify them slightly.

4) Word Mover's Distance (WMD)
   The WMD distance measures the dissimilarity between two text documents as the minimum amount of distance that the embedded words of one document need to "travel" to reach the embedded words of another document. The distance between word i and word j becomes $c(i, j) = \|V(i) - V(j)\|_2$.

3.3   The Feature Combination Method

Each feature set used in the paper is a combination of similarity metrics. Finding effective feature combinations is the key to improve the performance of the model. Traversing all feature combinations to find the optimal feature combination has a high time complexity. Therefore, a feature set generation algorithm based on pruning strategy is proposed. The detail is shown in Algorithm 1.

Algorithm 1. Feature set generation

Input:
  Two feature sets S_1 and S_2;
Output:
  Candidate feature sets $Candidates$ and corresponding F1-scores $Fscores$;
1: $Visited \leftarrow \emptyset$
2: $Candidates \leftarrow \emptyset$
3: $Fscores \leftarrow \emptyset$
4: **for** each $feature1 \in S_1$ **do**
5:   **for** each $feature2 \in S_2$ **do**
6:     $feature\_set \leftarrow \{feature1, feature2\}$
7:     $fscore \leftarrow$ **evaluate**($feature\_set$)
8:     $Candidates \leftarrow Candidates \cap \{feature\_set\}$
9:     $Fscores \leftarrow Fscores \cap \{fscore\}$
10:     $Visited \leftarrow Visited \cap \{feature\_set\}$
11:   **end for**
12: **end for**
13: $all\_features \leftarrow S_1 \cap S_2$
14: **for** $i = 0$ to $|all\_features| - 2$ **do**
15:   **for** each $feature\_set, fscore \in Candidates$ and $Fscores$ **do**
16:     **if** $|feature\_set| \neq 2 + i$ **then**
17:       **Continue**
18:     **end if**
19:     $left\_features \leftarrow all\_features - feature\_set$
20:     **for** each $feature \in left\_features$ **do**
21:       $new\_set \leftarrow feature\_set \cap feature$
22:       **if** $new\_set \in Visited$ **then**
23:         **Continue**
24:       **end if**
25:       $Visited \leftarrow Visited \cap \{new\_set\}$
26:       $new\_fscore \leftarrow$ **evaluate**(new_set)
27:       **if** $new\_fscore \geq fscore$ **then**
28:         $Candidates \leftarrow Candidates \cap \{new\_set\}$
29:         $Fscores \leftarrow Fscores \cap \{new\_fscore\}$
30:       **end if**
31:     **end for**
32:   **end for**
33: **end for**

Since containing more features does not ensure the improvement of performance, we propose a method to find the optimal combination of features. The feature set generation algorithm mainly contains two steps: initial feature set generation and feature set growth. The initial feature set is considered as candidate feature set. Growth is performed on these candidate feature sets based on a total of eight features obtained. If a new feature set obtained achieves better evaluation result, then the new feature set and corresponding evaluation result is added to the candidate feature set. In order to avoid duplication or useless attempts, each new feature set needs to be marked so that the growth of the candidate feature set can be pruned when the algorithm iterates.

The initial feature sets are obtained by cross-combining two types of similarity metrics. After that, their performances on the training set are evaluated, which are shown as lines 1–12. Improved feature sets can be found iteratively based on F-score. The loop statement on line 14 is used to iterate through all possible combinations of features. The statement on line 15 attempts to grow all candidate feature sets. If a new feature set achieves higher F-score, then the feature set is added to the candidate feature sets, otherwise it is discarded and will not be evaluated again, as shown in lines 22–30. Lines 22–25 implement a pruning strategy, ensuring that a feature set without improved F-score will not be tested again.

3.4 XGBoost Model

We used the XGBoost model proposed by Chen [3], which has been widely applied in different kinds of data mining tasks. XGBoost classifier is a boosting classifier which combines hundreds of tree models with lower classification accuracy into a stronger learner in an iterative fashion. At each iteration of gradient boosting, the residual is used to correct previous predictor such that the specified loss function can be optimized. Different from other Gradient Boosting Decision Tree (GBDT) model, regularization is added to the loss function to establish the objective function in XGBoost, which is given by:

$$J(\Theta) = L(\Theta) + \Omega(\Theta),$$

$$L(\Theta) = \sum_{i=1}^{n} l(\widehat{y_i}, y_i),$$

$$\Omega(\Theta) = \sum_{k=1}^{t} \tau T + \frac{1}{2}\lambda||w||^2$$

(7)

Here, Θ refers to the various parameters in the formula. $J(\Theta)$ is the objective function. $L(\Theta)$ is the training loss function that measures the difference between the prediction $\widehat{y_i}$ and the target y_i. Commonly used convex loss function such as square loss or logistic loss can be used in the above equation. $\Omega(\Theta)$ is a regularized term that penalizes complex models. In the definition of $\Omega(\Theta)$, T is the number of leaves in the tree, τ is the learning rate, λ is a regularized parameter to scale the penalty, and w is the weight of the leaves. Since the base model is decision tree, the result of prediction $\widehat{y_i}$ is the sum of scores predicted by K trees:

$$\widehat{y_i} = \sum_{k=1}^{t} f_k(x_i), f_k \in F$$

(8)

where x_i is the i-th training sample, F is the space of functions containing all regression trees and $f_k(x_i)$ is the score for the k-th tree. The optimization goal is to construct a tree structure that minimizes the target function in each iteration. The tree structure learns from conclusions and residuals of previous trees, fitting a current residual regression tree.

3.5 The Multi-model Fusion Method

In order to acquire better performance, a multi-model fusion method is proposed. Model fusion is an effective way to improve the accuracy of model, and voting weighted fusion is a fast and direct method. After feature sets are generated, the models obtained by different feature sets can be used for model fusion. The basic element of model fusion is to ensure that the correlation between individual models is small, and the performance difference between different models is not significant. Therefore, models obtained by different feature sets are used for model fusion.

The candidate feature sets are obtained through the feature set generation algorithm. After that, the first step is to generate multiple XGBoost models with different feature sets as input, and then calculate the value of F1 for each model respectively. Let s_i represents a collection of feature sets consisting of i features, p_i is the selected feature set. Then

$$p_i = \arg\max_{j \in s_i} fscore(j) \tag{9}$$

The second step is the selection of the models. Each of these models is an XGBoost model trained by an optimal feature set of a specific number of features. The model fusion method is to conduct weighted voting according to the F1 score of the selected model. Assume that N' models are selected, the final prediction of sample i is:

$$\widehat{y_i} = \sum_{j=1}^{N'} \widehat{y_i}(j) w_j \tag{10}$$

In this formula, $\widehat{y_i}(j)$ is prediction for sample i given by model j, w_j is the voting weight of the model j. Models with higher F1 scores usually have better classification performance, so w is calculated based on the F1 score of the model on the training set. For $j \in [1, N']$:

$$w_j = \frac{fscore(j)}{\sum_{i=1}^{N'} fscore(i)} \tag{11}$$

4   Evaluation and Results

4.1   Datasets

We applied the same QA dataset as used in existing baseline methods, e.g., [6]. The dataset was generated from 25569 questions groups that shared the same answer extracted in Yahoo! Answers. We filtered the answers that were too short, e.g, "*Yeah*", and obtained a total of 624 question groups. Eventually, the dataset contained questions with a maximum length of 25 words, a minimum length of 2 words, an average length of 10 words, and a standard deviation of 4.50.

We divided the dataset into a training set containing 524 question groups and a testing set containing 100 question groups randomly, similar to the baseline method. Since the number of samples in the dataset was relatively small, the training results fluctuated greatly by different division methods. The experiment were randomly repeated 100 times and the average of the evaluation results was considered.

4.2   Baselines

The baseline methods for comparison were widely used text similarity calculation measures as follows:

1) Topical Diversity: We used TD+POS4 method described in Sect. 3.2.
2) Jaccard: Described in Sect. 3.2.
3) CosBERT: Single question similarity detection model using feature described in Sect. 3.2.
4) WMD: Unsupervised method to calculated the semantic distance between two questions using Word2Vec model. Described in Sect. 3.2.
5) CosSIF: Described in Sect. 3.2.
6) CosW2V: Described in Sect. 3.2.
7) Metric Longest Common Subsequence: The ratio of the number of words in the longest subsequence of two questions to the number of words in the longest question.
8) NN: A multi-feature fusion method based on neural network proposed by Ruan et al. [18].

Feature extraction involves two pre-training models with training parameters and sources as follows. Word2Vec [13] was trained on Google News dataset, which contains 300-dimensional vectors for 3 million words and phrases. The resulting vectors have been made publicly available[3]. BERT [5] was multilingual uncased BERT-Base model[4]. The model gives 768-dimensional vectors for each word.

In the experiments, we used the same model configuration for model training of each feature set. The learning rate and was set to 0.05. The maximum depth of tree in each booster and was set to 3. Evaluation metric was a logarithmic loss function. These parameters were set empirically.

4.3   Results

Four experiments were conducted to evaluate the effectiveness of the proposed method for similar question detection. The evaluation metrics were precision, recall, and F1 score, which were commonly used in information retrieval area.

---

[3] see https://code.google.com/archive/p/word2vec/.
[4] see https://github.com/google-research/bert.

Table 1. The performance comparison with the different feature sets (%)

Features	Methods	Avg precision	Avg recall	Avg F1
TWM, Jac, CosSIF	Random Forest	84.52	70.98	76.43
CosBERT, TD, Jac, CosSIF		83.61	71.31	76.32
Jac, CosSIF		82.89	71.88	76.34
TWM, Jac, CosSIF	SVM	93.20	53.63	67.42
CosBERT, TD, Jac, CosSIF		**94.28**	53.50	67.56
Jac, CosSIF		91.94	55.59	68.67
TWM, Jac, CosSIF	XGBoost	85.37	72.64	**77.99**
CosBERT, TD, Jac, CosSIF		85.30	**72.74**	77.97
Jac, CosSIF		83.34	72.22	76.77

First, an experiment to evaluate the performance of feature combinations was conducted. According to the feature set generation algorithm, candidate feature sets sorted according to F1 score can be obtained. Then the feature sets with the highest F1 score of two, three and four features were selected, namely FS1(Jac + CosSIF), FS2(TWM + Jac + CosSIF) and FS3(CosBERT + TD + Jac + CosSIF). These three feature sets were then used to train different classifiers, including SVM, Random Forest, and XGBoost, and compare their performances. The result showed that the feature set FS2 achieved the highest F1 in the XGBoost classifier. The feature set with the largest number of features in candidate sets has five features, but the F1 score is lower than that of the feature set with only two features in SVM and Random Forest. The result is shown as Table 1. From the result, the XGBoost model with TWM, Jac and CosSIF as feature set achieves the highest average F1.

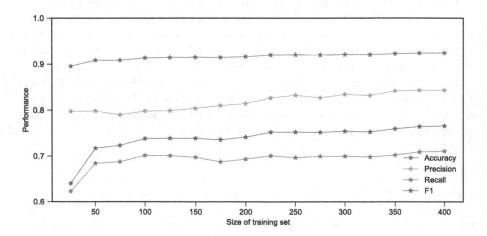

Fig. 2. The performance of fusion XGBoost training using different sizes of data

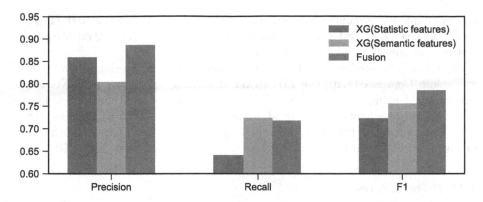

Fig. 3. The performance in datasets using different types of features

The second experiment evaluates how question similarity calculation performance was affected by the size of training datasets to test the scalability of our model. The used datasets were the same as used in the first experiment.

As shown in Fig. 2, when the number of samples in the training set was more than 225, the performance of the model became stable. When the number of samples was less than 225, the performance of the model fluctuated significantly. This demonstrated that the model was effective when the number of training set samples was more than 225.

We compared the performance obtained by single XGBoost model using two different types of features as input. The XG (Statistic features) model using all features described in our statistic similarity metrics, and XG (Semantic features) using all semantic similarity metrics. The Fusion model using our proposed fea-

Table 2. The performance comparison with baselines (%)

Methods	Avg accuracy	Avg precision	Avg recall	Avg F1
CosSIF	87.17	60.72	70.91	64.95
Metric Longest Common Subsequence	89.64	70.45	67.58	68.49
WMD	90.41	74.89	70.57	70.89
CosW2V	90.65	75.02	69.17	71.02
CosBERT	90.37	71.79	72.13	71.49
Jaccard	91.05	77.39	67.68	71.63
Topical Diversity (TD)	91.33	82.07	65.10	71.49
NN	62.15	48.75	**76.68**	51.40
XG (TWM+Jaccard+CosSIF)	93.17	85.37	72.64	77.99
XG (CosBERT+TD+Jaccard+CosSIF)	93.14	85.30	72.74	77.97
XG (Jaccard+CosSIF)	92.74	83.34	72.22	76.77
Fusion (FS1+FS2)	92.80	**89.56**	69.49	77.42
Fusion (FS1+FS2+FS3)	**94.60**	88.65	71.85	**78.54**

ture combination and multi-model fusion method. According to Fig. 3, the Fusion model achieved highest F1, which demonstrated the effectiveness of our proposed feature combination and multi-model fusion method.

We then compared our method with baseline methods. The result was shown in Table 2. Compared with the TD+POS4 algorithm, our approach of XGBoost single model using the feature group TWM+Jaccard+CosSIF achieves 4.02% precision improvement, 11.58% recall improvement and 9.09% F1 improvement. The fusion model of the three different length feature sets reached the highest precision and F1, and F1 is increased by 9.86% compared with TD+POS4.

4.4   Error Analysis

The result errors including undetected and incorrectly identified similar questions in the experiments were collected and systematically analyzed. Since the models were evaluated by randomly dividing the data set 100 times, we calculated the cases of prediction errors in 100 tests using the fusion model FS1 + FS2 + FS3.

16 groups of questions were predicted to be incorrect more than 10 times, and a total of 5 groups of questions were detected by our method. We found it is difficult for our model to detect similar question pairs with uncommon words, non-generic words, and abbreviations, e.g., PSP and SUV. Two samples of this type of error are shown in Table 3.

Table 3. Examples of undetected error by our method

Group ID	Questions	Answer
G1	*Where can i find free downloads for my PSP?*	*www.yourpsp.com*
	I saw an advert for a psp related site for downloads it had 2 dust balls or something like that whats the site	
G2	*What is SUV?*	*Sports utilize vehicle*
	SUV meaning?	

11 groups of questions that were incorrectly judged to be similar by our model. There was a common denominator of this type of error, where one question was much longer than the other and the longer one contained the same or related words of the short questions. For instance, the question *"Does anyone know this logo?"* and the question *"anybody know what the company logo is with the singing parrot?"* were marked as similar questions but were annotated dissimilar by human annotators. By analyzing this type of errors, we found that some question pairs had a high proportion of identical or related words, resulting in close semantic features. Therefore, the question target or answer type may be a key factor for eliminating the effect of the word overlap.

5   Conclusion

In this paper, we have applied a fusion XGBoost model to detect similar questions. A feature set generation and a voting method based on F1 score of each model are proposed. Experiments on Yahoo! dataset have showed that the proposed method is feasible in improving the performance of similar question detection. In the future, we plan to evaluate the proposed method on more datasets and consider more features of questions.

Acknowledgement. This work was supported by National Natural Science Foundation of China (No.61772146), Natural Science Foundation of Guangdong Province (2018A030310051), and the Katie Shu Sui Pui Charitable Trust – Research and Publication Fund (KS 2018/2.8).

References

1. Arora, S., Liang, Y., Ma, T.: A simple but tough-to-beat baseline for sentence embeddings (2016)
2. Chali, Y., Islam, R.: Question-question similarity in online forums. In: The 10th Annual Meeting of the Forum for Information Retrieval Evaluation, pp. 21–28. ACM (2018)
3. Chen, T., Guestrin, C.: XGBoost: a scalable tree boosting system. In: The 22nd ACM SIGKDD International Conference on Knowledge Discovery and Data Mining, pp. 785–794. ACM (2016)
4. Das, A., Shrivastava, M., Chinnakotla, M.: Mirror on the wall: finding similar questions with deep structured topic modeling. In: Bailey, J., Khan, L., Washio, T., Dobbie, G., Huang, J., Wang, R. (eds.) PAKDD 2016. LNCS, vol. 9652, pp. 454–465. Springer, Cham (2016). https://doi.org/10.1007/978-3-319-31750-2_36
5. Devlin, J., Chang, M.W., Lee, K., Toutanova, K.: Bert: pre-training of deep bidirectional transformers for language understanding. arXiv:1810.04805 (2018)
6. Hao, T., Li, C., Liang, W., Qu, Y.: A topical diversity-based approach to detecting similar question groups from collaborative question-answering archives. In: Web Intelligence, vol. 14, pp. 301–308. IOS Press (2016)
7. Heeringa, W.J.: Measuring dialect pronunciation differences using Levenshtein distance. Ph.D. thesis. Citeseer (2004)
8. Huang, C.H., Yin, J., Hou, F.: A text similarity measurement combining word semantic information with TF-IDF method. Chin. J. Comput. **34**(5), 856–864 (2011)
9. Jiang, J.J., Conrath, D.W.: Semantic similarity based on corpus statistics and lexical taxonomy. arXiv preprint cmp-lg/9709008 (1997)
10. Kusner, M., Sun, Y., Kolkin, N., Weinberger, K.: From word embeddings to document distances. In: ICML, pp. 957–966 (2015)
11. Li, S., Zhang, J., Huang, X., Bai, S., Liu, Q.: Semantic computation in a Chinese question-answering system. J. Comput. Sci. Technol. **17**(6), 933–939 (2002)
12. Li, W., Zhao, Y.: Semantic similarity between concepts algorithm based on ontology structure. Jisuanji Gongcheng/ Comput. Eng. **36**(23) (2010)
13. Mikolov, T., Sutskever, I., Chen, K., Corrado, G.S., Dean, J.: Distributed representations of words and phrases and their compositionality. In: Advances in Neural Information Processing Systems, pp. 3111–3119 (2013)

14. Nakov, P., et al.: SemEval-2017 task 3: community question answering. arXiv preprint arXiv:1912.00730 (2019)
15. Niwattanakul, S., Singthongchai, J., Naenudorn, E., Wanapu, S.: Using of Jaccard coefficient for keywords similarity. In: The International Multiconference of Engineers and Computer Scientists, vol. 1, pp. 380–384 (2013)
16. Qiu, X., Huang, X.: Convolutional neural tensor network architecture for community-based question answering. In: IJCAI (2015)
17. Ramos, J., et al.: Using TF-IDF to determine word relevance in document queries. In: The First Instructional Conference on Machine Learning, Piscataway, NJ, vol. 242, pp. 133–142 (2003)
18. Ruan, H., Li, Y., Wang, Q., Liu, Y.: A research on sentence similarity for question answering system based on multi-feature fusion. In: 2016 IEEE/WIC/ACM International Conference on Web Intelligence (WI), pp. 507–510. IEEE (2016)
19. Song, W., Feng, M., Gu, N., Wenyin, L.: Question similarity calculation for FAQ answering. In: SKG, pp. 298–301. IEEE (2007)
20. Uva, A., Bonadiman, D., Moschitti, A.: Injecting relational structural representation in neural networks for question similarity. In: ACL, pp. 285–291 (2018)
21. Wang, K., Ming, Z., Chua, T.S.: A syntactic tree matching approach to finding similar questions in community-based QA services. In: The 32nd International ACM SIGIR Conference on Research and Development in Information Retrieval, pp. 187–194. ACM (2009)
22. Ye, B., Feng, G., Cui, A., Li, M.: Learning question similarity with recurrent neural networks. In: 2017 IEEE ICBK, pp. 111–118. IEEE (2017)
23. Zhou, G., Cai, L., Zhao, J., Liu, K.: Phrase-based translation model for question retrieval in community question answer archives. In: ACL, pp. 653–662 (2011)
24. Zhou, G., Zhou, Y., He, T., Wu, W.: Learning semantic representation with neural networks for community question answering retrieval. Knowl.-Based Syst. **93**, 75–83 (2016)

Core Research Topics of Studies on Personalized Feedback in the Past Four Decades

Xieling Chen[1], Di Zou[2(✉)], Gary Cheng[1], Haoran Xie[3], Fu Lee Wang[4], and Leung Pun Wong[4]

[1] Department of Mathematics and Information Technology, The Education University of Hong Kong, Hong Kong, Hong Kong SAR

[2] Department of English Language Education, The Education University of Hong Kong, Hong Kong, Hong Kong SAR
dizoudaisy@gmail.com

[3] Department of Computing and Decision Sciences, Lingnan University, Hong Kong, Hong Kong SAR

[4] School of Science and Technology, The Open University of Hong Kong, Hong Kong, Hong Kong SAR

Abstract. Assessment feedback is an essential part of learners' learning experiences. Personalized feedback in learning is a useful and common strategy for assisting learners to optimize their learning. With the increasing need to provide learners with high quality, immediate, and personalized feedback, a large number of studies had been conducted to investigate how to provide students with personalized feedback effectively. In this study, bibliometric analysis and word cloud techniques were applied to identify research trends and status related to personalized feedback in teaching and learning, based on 276 publications retrieved from the Web of Science database. To be specific, the data were analyzed in terms of annual numbers of publications and citations, important publication sources, countries/regions, and institutions, as well as important research issues and concerns. The findings of this study provided scholars as well as instructors with a general picture of the personalized feedback research.

Keywords: Personalized feedback · Bibliometric analysis · Word cloud

1 Introduction

Feedback refers to specific information concerning a learners' performance regarding a defined standard, given with intention for performance improvement [1]. Formative feedback is important for teaching and learning, especially for online programs. For a great number of courses, feedbacks are given by instructors, demanding high quality in content and process, particularly in introductory and basically procedural courses. Feedback may assume various forms, and the effectiveness and appropriacy of various feedback approaches differ [2]. Feedback is an essential component of assessment in learning contexts, enabling learners to monitor progress in the learning process, and helping instructors to provide personalized learning materials based on learners' profiles

© Springer Nature Switzerland AG 2020
Y. Nah et al. (Eds.): DASFAA 2020 Workshops, LNCS 12115, pp. 121–130, 2020.
https://doi.org/10.1007/978-3-030-59413-8_10

[3]. With the increasing popularity of personalized learning (e.g., [4–7]), personalized feedback has become increasingly important [8].

A growing number of automatic models allow making inferences about learners' understanding according to their problem-solving choices, with various applications such as personalized feedback interventions in interactive educational contexts [9]. Studies had suggested that automated, computer-generated feedback allowed addressing learners' need for receiving feedback, as well as instructors' need to offer useful feedback efficiently [10]. Personalized feedback allows transforming self-assessment experience into learners' learning experience [3], and the personalization of feedback has become an essential research issue in electronic learning systems. Scholars have illustrated that the merits of computer-generated feedback were personalized and teacher-generated [10, 11], and highlighted that electronic feedback was more effective as compared to traditional methods [12].

Scholars have been devoting to research on feedback to allow self-assessment by using information generated during the learning process and further to offer personalized feedback to individual learners [3]. Currently, no review has been conducted in this field. To that end, this study aimed to explore research status and trends of personalized feedback using bibliometric analysis and word cloud technique. As effective and useful methods in mapping academic literature, bibliometrics and word cloud have been popularly adopted in many fields of research [13–18]. Based on 276 publications retrieved from Web of Science database, we analyzed the annual numbers of publications and citations, relevant publication sources, countries/regions, and institutions, as well as essential research issues and concerns.

2 Data and Methods

Figure 1 shows the workflow of data collection and analysis, including steps of data retrieval, data restriction, manual screening, and data analysis. The data retrieval was carried out in Web of Science database on February 26, 2020, with a search query written as TS = (("personalized" OR "personalised" OR "personalisation" OR "personalization" OR "personalizing" OR "personalising") AND "feedback"). The initial search returned 3,294 publications. The data were further narrowed down using the following restrictions to make sure that the publications were: 1) written in English, 2) research articles or conference papers, and 3) Education and Educational research. In this way, 385 publications were selected. Two domain experts then screened the 385 publications manually to ensure that they were closely related to personalized feedback for teaching and learning purposes, with irrelevant ones being excluded, including: 1) reviews (n = 12), 2) irrelevant to teaching and learning (n = 17), 3) surveys (n = 5), 4) irrelevant to personalized feedback (n = 70), and 5) published in 2020 (n = 5). After screening, 267 publications remained for data analysis.

To explore the annul trends of publications and citations, we carried out polynomial regression analyses with *year* as independent variable *x*. Analyses of important publication sources, countries/regions, as well as institutions, were conducted based on several bibliometric indicators, including publication count, citation count, average citations per publication, as well as Hirsh index (H-index) [19]. As for the analysis of research topics, key phrases extracted from the title and abstract of each publication were utilized. After pre-processing, as suggested by Chen et al. [13], the key phrases were analyzed using word cloud technique using R package *wordcloud2*. Three consecutive time periods were used to explore the evolution of essential phrases, including 1981–2004 (11 publications), 2005–2014 (114 publications), and 2015–2019 (151 publications).

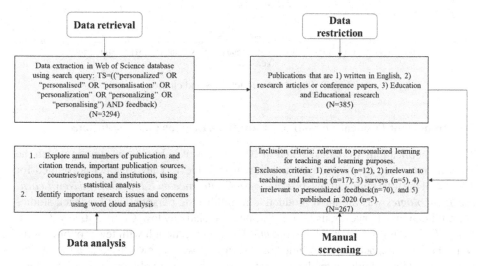

Fig. 1. The workflow of data collection and analysis

3   Results and Discussions

3.1   Trends of Publications and Citations

The trends of personalized feedback publications and the citations they received are depicted in Fig. 2, together with their polynomial regression curves. From the results, it was clear that for both the citations and publications, significantly increasing trends were shown. The results could also be indicated from the two regression models with positive coefficients of x^2. With the two estimated models, predictive values for future years could be estimated. For example, the predictive values for the year 2020 in terms of the publication and citation counts were calculated as $0.04455176*2020^2 - 177.4491*2020 + 176692.7 = 34.4123$ and $0.3285617*2020^2 - 1309.215*2020 + 1304186 = 234.904$. In a word, the trends and the regression modeling results demonstrated a growing interest in the research on personalized feedback for teaching and learning, which is an increasingly important and impactful field of research.

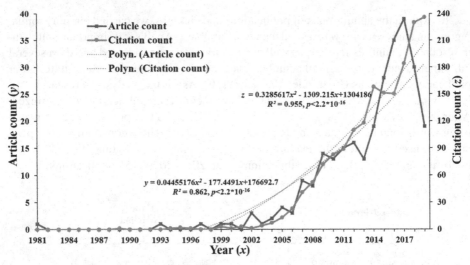

$$z = 0.3285617x^2 - 1309.215x + 1304186$$
$$R^2 = 0.955, p < 2.2*10^{-16}$$

$$y = 0.04455176x^2 - 177.4491x + 176692.7$$
$$R^2 = 0.862, p < 2.2*10^{-16}$$

Fig. 2. Trends of publications and citations

3.2 Important Publication Sources, Countries/Regions and Institutions

The 276 publications were distributed in 106 sources of publications, among which the top nine measured by productivity are shown in Table 1. They together accounted for 40.22% of the total publications. *International Conference on Education and New Learning Technologies* was the most productive in publishing personalized feedback studies (32 publications). However, its H-index value was relatively low. Comparatively, *Educational Technology & Society* and *Academic Medicine*, although with fewer publications, were the most impactful in the field. In addition, it was noteworthy that although the top six listed in the table were all conferences, they had relatively lower H-index values. Thus, we might conclude that research articles in the field of personalized feedback were more impactful than conference papers.

Table 1. Top publication sources

Publication sources	1981–2019			1981–2004		2005–2014		2015–2019	
	A	C	H	A	C	A	C	A	C
International Conference on Education and New Learning Technologies	32	6	1	0	0	13	0	19	6
International Conference of Education, Research and Innovation	21	4	1	0	0	8	1	13	3
International Technology, Education and Development Conference	14	5	1	0	0	7	0	7	5
European Conference on E-learning	9	8	2	0	0	9	3	0	5
Frontiers in Education Conference	9	0	0	0	0	3	0	6	0
European Conference on Games Based Learning	8	9	2	0	0	5	1	3	8
Academic Medicine	6	221	4	0	0	3	159	3	62
Educational Technology & Society	6	109	4	0	0	3	58	3	51
International Conference on E-learning	6	5	1	0	0	4	0	2	5

Abbreviations: H: H-index; A: publication count; C: citation count.

A total of 58 countries/regions had contributed to the 276 publications. Table 2 lists the top ten prolific countries/regions, among which the USA was ranked the first as measured by publication count, citation count, and H-index (62 publications, 684 citations, and an H-index of 11). Using the measure of average citations per publication (ACP), it was noteworthy that although with relatively fewer publications, Taiwan was ranked at the top with an ACP value of 38.78. This indicated the broad influence of Taiwan's publications in personalized feedback.

Table 2. Top countries/regions

Countries/regions	1981–2019				1981–2004		2005–2014		2015–2019	
	A	C	ACP	H	A	C	A	C	A	C
USA	62	684	11.03	11	6	14	25	360	31	310
UK	42	165	3.93	8	1	5	20	29	21	131
China	22	23	1.05	3	0	0	5	1	17	22
Spain	22	13	0.59	2	0	0	12	1	10	12
Romania	15	18	1.20	2	0	0	6	8	9	10
Australia	13	71	5.46	5	2	2	3	6	8	63
Canada	12	145	12.08	6	1	2	7	52	4	91
Greece	10	40	4.00	3	0	0	8	24	2	16
India	9	52	5.78	2	0	0	3	7	6	45
Taiwan	9	349	38.78	6	0	0	5	191	4	158

Abbreviations: H: H-index; A: publication count; C: citation count; ACP: citation count per publication.

A total of 325 institutions had contributed to the 276 publications. Table 3 lists the top eight prolific institutions, among which *University of Barcelona* was the top one in terms of publication count. However, its citation count and H-index value were relatively lower, particularly as compared to *Athabasca University* (5 publications, 78 citations, and an H-index of 4). From the perspective of ACP, *Athabasca University* was ranked at the top one (15.60). Taking into account of all the indicators, we concluded that *Athabasca University* could be regarded as the top contributor to the research on personalized feedback, particularly in terms of research impact and influence.

Table 3. Top institutions

Institutions	Countries/regions	1981–2019				1981–2004		2005–2014		2015–2019	
		A	C	H	ACP	A	C	A	C	A	C
University of Barcelona	Spain	6	2	1	0.33	0	0	4	0	2	2
Athabasca University	Canada	5	78	4	15.60	0	0	3	20	2	58
University Politehnica of Bucharest	Romania	5	1	1	0.20	0	0	1	0	4	1
The Bucharest University of Economic Studies	Romania	4	11	1	2.75	0	0	1	6	3	5
The Open University	UK	4	15	1	3.75	0	0	1	0	3	15
University of Leeds	UK	4	52	2	13.00	0	0	3	11	1	41
Universitat Oberta de Catalunya	Spain	4	3	1	0.75	0	0	3	0	1	3
University of Sydney	Australia	4	18	2	4.50	0	0	1	0	3	18

Abbreviations: H: H-index; A: publication count; C: citation count; ACP: citation count per publication.

3.3 Important Research Topics

Table 4 lists the top 15 frequently used phrases in personalized feedback studies, among which "personalized feedback (appearing in 54 publications, accounting for 19.57%)" ranked at the first, followed by "learning process (46, 16.67%)," "personalized learning (29, 10.52%)," "learning experience (26, 9.42%)," "learning outcome (21, 7.61%)," "online learning (19, 6.88%)," and "virtual learning environment (19, 6.88%)." From the evolutions of the important issues for the three periods, as shown in both Table 4 and Fig. 3, some research issues had become increasingly concerned by scholars. First, several issues had been studied more over time, for example, "learning process," "personalized learning," "online learning," "formative assessment," "personalized feedback," and "learning environment." Second, there were several issues emerged in the latter two periods, for example, "learning experience," "learning outcome," "virtual learning environment," "case study," "e-learning system," "learning activity," "learning management system," and "learning system." In addition, "learning analytics" appeared to be important in the last period, indicating its wide and popular application in personalized feedback research.

Table 4. Top frequently used phrases

Key phases	1981–2019		1981–2004		2005–2014		2015–2019	
	A	%	A	%	A	%	A	%
Personalized feedback	54	19.57%	4	36.36%	23	20.18%	27	17.88%
Learning process	46	16.67%	1	9.09%	25	21.93%	20	13.25%
Personalized learning	29	10.51%	1	9.09%	8	7.02%	20	13.25%
Learning experience	26	9.42%	0	0.00%	12	10.53%	14	9.27%
Learning outcome	21	7.61%	0	0.00%	8	7.02%	13	8.61%
Online learning	19	6.88%	1	9.09%	6	5.26%	12	7.95%
Virtual learning environment	19	6.88%	0	0.00%	11	9.65%	8	5.30%
Formative assessment	17	6.16%	1	9.09%	7	6.14%	9	5.96%
Learning analytics	17	6.16%	0	0.00%	0	0.00%	17	11.26%
Case study	16	5.80%	0	0.00%	7	6.14%	9	5.96%
E-learning system	16	5.80%	0	0.00%	13	11.40%	3	1.99%
Learning activity	16	5.80%	0	0.00%	8	7.02%	8	5.30%
Learning environment	16	5.80%	1	9.09%	8	7.02%	7	4.64%
Learning management system	16	5.80%	0	0.00%	7	6.14%	9	5.96%
Learning system	16	5.80%	0	0.00%	8	7.02%	8	5.30%

Abbreviations: A: publication count; %: percentage of publications.

Fig. 3. Term evolution during the periods 1981–2004, 2005–2014, and 2015–2019.

4   Conclusion

In general terms, within the bibliometric analysis, there was evidence of an increase in research interest and scientific impact of personalized feedback for teaching and learning. Several growing trends had been identified. First, studies on the personalized feedback in web- or online-based learning environment had become more and more popular among scholars. Second, studies concerning the provision of personalized feedback based on learners' learning processes, learning experiences, and learning outcomes in learning activities were increasingly available. In addition, studies adopting learning analytics for providing personalized feedback were particularly concerned by scholars, especially in recent years.

Acknowledgments. This research received grants from the Standing Committee on Language Education and Research (EDB(LE)/P&R/EL/175/2), the Education Bureau of the Hong Kong Special Administrative Region, the Internal Research Grant (RG93/2018–2019R), and the Internal Research Fund (RG 1/2019–2020R), the Interdisciplinary Research Scheme of the Dean's Research Fund 2018–19 (FLASS/DRF/IDS-3) and the Small Research Grant for Academic Staff 2019–20 (MIT/SGA04/19–20) of The Education University of Hong Kong.

References

1. Van De Ridder, J.M., Stokking, K.M., McGaghie, W.C., Ten Cate, O.T.J.: What is feedback in clinical education? Med. Educ. **42**, 189–197 (2008)
2. Belcadhi, L.C.: Personalized feedback for self assessment in lifelong learning environments based on semantic web. Comput. Hum. Behav. **55**, 562–570 (2016)
3. Xie, H., Chu, H.C., Hwang, G.J., Wang, C.C.: Trends and development in technology-enhanced adaptive/personalized learning: a systematic review of journal publications from 2007 to 2017. Comput. Educ. **140**, 103599 (2019)
4. Xie, H., Zou, D., Zhang, R., Wang, M., Kwan, R.: Personalized word learning for university students: a profile-based method for e-learning systems. J. Comput. High. Educ. **31**, 273–289 (2019)

5. Zou, D., Xie, H.: Personalized word-learning based on technique feature analysis and learning analytics. J. Educ. Technol. Soc. **21**, 233–244 (2018)
6. Xie, H., Zou, D., Lau, R.Y., Wang, F.L., Wong, T.L.: Generating incidental word-learning tasks via topic-based and load-based profiles. IEEE Multimed. **23**, 60–70 (2015)
7. Planar, D., Moya, S.: The effectiveness of instructor personalized and formative feedback provided by instructor in an online setting: some unresolved issues. Electron. J. E-Learn. **14**, 196–203 (2016)
8. Rafferty, A.N., Jansen, R., Griffiths, T.L.: Using inverse planning for personalized feedback. EDM **16**, 472–477 (2016)
9. Ene, E., Kosobucki, V.: Rubrics and corrective feedback in ESL writing: a longitudinal case study of an L2 writer. Assess. Writ. **30**, 3–20 (2016)
10. Stevenson, M., Phakiti, A.: The effects of computer-generated feedback on the quality of writing. Assess. Writ. **19**, 51–65 (2014)
11. Mayen, P., Savoyant, A.: Formation et prescription: une réflexion de didactique profes-sionnelle. In: Revest, C., Schwartz, Y. (eds.) Les évolutions de la prescription-Actes du XXXVIIème congrès de la SELF. Toulouse, Octarès Editions, pp. 226–232 (2002)
12. Chen, X., Zou, D., Xie, H.: Fifty years of British journal of educational technology: a topic modeling based bibliometric perspective. Br. J. Educ. Technol. **51**(3), 692–708 (2020)
13. Song, Y., Chen, X., Hao, T., Liu, Z., Lan, Z.: Exploring two decades of research on classroom dialogue by using bibliometric analysis. Comput. Educ. **137**, 12–31 (2019)
14. Chen, X., Zou, D., Cheng, G., Xie, H.: Detecting latent topics and trends in educational technologies over four decades using structural topic modeling: a retrospective of all volumes of computer & education. Comput. Educ. **151**, 103855 (2020)
15. Chen, X., Xie, H., Wang, F.L., Liu, Z., Xu, J., Hao, T.: A bibliometric analysis of natural language processing in medical research. BMC Med. Inform. Decis. Mak. **18**, 14 (2018)
16. Chen, X., Yu, G., Cheng, G., Hao, T.: Research topics, author profiles, and collaboration networks in the top-ranked journal on educational technology over the past 40 years: a bib-liometric analysis. J. Comput. Educ. **6**(4), 563–585 (2019). https://doi.org/10.1007/s40692-019-00149-1
17. Chen, X., Wang, S., Tang, Y., Hao, T.: A bibliometric analysis of event detection in social media. Online Inf. Rev. **43**, 29–52 (2019)
18. Hirsch, J.E., Buela-Casal, G.: The meaning of the h-index. Int. J. Clin. Health Psychol. **14**, 161–164 (2014)
19. Zou, D., Lambert, J.: Feedback methods for student voice in the digital age. Br. J. Educ. Technol. **48**, 1081–1091 (2017)

Maintenance Method of Logistics Vehicle Based on Data Science and Quality

Yu Ping Peng[1], Shih Chieh Cheng[2(✉)], Yu Tsung Huang[2], and Jun Der Leu[2]

[1] Department of Logistics Management, National Defense University, Taoyuan City, Taiwan
jason@taivs.tp.edu.com
[2] Department of Business Administration, National Central University, Taoyuan City, Taiwan
leujunder@mgt.ncu.edu.tw

Abstract. With the changes in consumption around the world, the global logistics and logistics management has been developed, which has derived the business opportunities of freight logistics and the demand for vehicles, which has led to the increase of vehicles service and service parts . Therefore, effective remaining vehicle readiness and reduction in maintain costs have become an urging subject to be solved in the industry of today. However, as in the era of big data, it is important that the enterprise makes good use of data and information to save costs, increase revenue, and ensure competitive advantages. But if we could not ensure the quality of the data, it would easily lead the analysis to the wrong decisions. Therefore, this study is based on the predictive maintenance, taking the condition of data quality considerations, and using algorithms to construct a decision support model, and proposing optimal replacement cycles and rules for vehicle components, and analyzing the impact on brands and maintenance amounts. Therefore, this study is based on maintenance history, through systematic and manual analysis, to obtain good quality data, and then use chi-square test and algorithm analysis to establish a classification model for decision support. The research department analyzes and analyzes the 3.5 ton freight vehicle maintenance and repair history of a case company from 2008–2016. After the data is cleaned and sorted, it obtains 173,693 work orders and good data quality data for 23 types of maintenance items. And the results show that: the costs contains significant divergence among brands; service parts damage is related to particular environment; we can obtain appropriate service period through proper classification rules. The decision support model constructed by this study will be improved and integrated with the actual needs of the industry on the premise of taking into account the quality of data.

Keywords: Big data · Data quality · Vehicle predictive maintenance

1 Introduction

In recent years, the popularity of Internet has led to the changes of consumption and the booming of e-commerce, and one of the main causes is commodity delivery which has derived the enormous business opportunity of freight logistics and the demand for vehicles, then rise to huge business opportunities in the logistics and freight and demand,

© Springer Nature Switzerland AG 2020
Y. Nah et al. (Eds.): DASFAA 2020 Workshops, LNCS 12115, pp. 131–145, 2020.
https://doi.org/10.1007/978-3-030-59413-8_11

which has gradually become one of the key development projects of industry in many countries [10, 16]. As being in the era of big data, the data and information has been exploded growth of every enterprise. Just as the logistics industry has increased its demand for vehicles as its business has grown, the maintenance of its vehicles and the wear and tear of its components have also grown simultaneously. For example, the Hsinchu Transport company has spent nearly 8 million every month in 2008 for each type of vehicle maintenance costs [33]. As the core tool of urban logistics is a vehicle, reducing vehicle maintenance costs and extending the service life are important issues that the industry needs to resolve. It is a huge and complicate data of vehicle maintenance, and it is an Operation Focus of an enterprise that how to make sure the correction and quality of data and mining, analyzing the information and knowledge behind the data [22] to ensure the vehicle has proper maintenance to maintain the work effectiveness. Because of the source and quantity of vehicle history and maintenance data were generated from different departments in enterprise, so we need effective methods to extract useful information from the complicated and huge data; data mining technology, which has been widely used in analysis of large amounts of data, is especially suitable for knowledge mining work [6]. And in the relevant researches of logistics transportation, and the maintenance and replacement rules in different fields were rarely discussed. With fewer systems and manual methods, after the data is sorted and cleaned, the data with better data quality will be analyzed and calculated.

Based on this, the purpose of this research is to verify the quality of the data by using the system and manual methods of the company's vehicle component repair and maintenance data. Then use good data to verify whether there are differences in maintenance costs between vehicle attributes and brands. And by understanding the rules of maintenance and replacement in different factories and regions, the diagnosis and decision-making of predictive maintenance of vehicle components is put forward. In order to prevent the breakdown of the vehicle during the operation, it can obtain the optimal maintenance timing. The research framework is as follows: in Sect. 2, the data quality, data mining and forms of maintenance and other literature will be described; Sect. 3 is about case study and methodology; Sect. 4 is about the data quality validation and description of the analysis of each item of data; Sect. 5 is about the discussion about the aforementioned validation; in last section Sect. 6, the findings and recommendations obtained in this study will be described.

2   Literature Review

2.1   Data Quality and Data Mining

In the age of wisdom economy and big data, data has become an important source of growth and development of the government and enterprises, but too much data cannot be used effectively but become a heavy *burden for enterprises and organizations.* Therefore, to convert data into useful information and knowledge as a basis for decision support reference has been apparently become an important direction of business decisions at the present time [9]. However, big data sources have a wide range of complex structures. Data often comes from the interaction and integration of different systems and periods in the organization; Therefore, the acquired data may have related quality problems such as data error, information loss, inconsistency, and noise. Therefore, data cleaning has become one of the important methods for obtaining high-quality data. The most important purpose of the so-called data cleaning or quality control is to detect and eliminate errors, inapplicability and inconsistencies in the data in order to improve its data quality [22]. However, this method often deletes 20% to 40% of the data. Therefore, in order to save data, organizations or enterprises often supplement the missing value in the original data directly through long-term data trends. In practice, this approach may increase the uncertainty of the data, and even lead to the use of problematic data or replacement, which may cause doubts about the gap [21]. Therefore, the way data is cleaned is even more important. Wang et al. [32] proposed that data cleansing can be divided into four modes: manual execution, writing special applications, data cleansing that has nothing to do with specific application areas, and solving one type of problem. This study uses manual data cleanup and other methods to clean up the data.

The good data quality data obtained after cleaning can be used for data mining. Fayyad et al. [15] proposed the so-called data mining means to have a data analysis through algorithms, and then to find out a specific pattern or model, which is also the procedure how to discover knowledge from data base. Berry and Linoff et al. [5] believed that data mining is to analyze a great amount of data in automated or half-automated ways and then from which to find out meaningful relationships or rules. To sum up, the so-called data mining is a method to derive a way of obtaining decision-making relevant information by analyzing a large number of data [27]. And today the data mining has been widely used in manufacturing, finance, service, medical and marketing and other different industries. But when it is used in the field of logistics, such as the aforementioned, Only a few researches focused on relevant issues about the maintenance of vehicles. Therefore, this study is to discuss the implementation of the logistics methods of vehicle maintenance in freight logistics and mainly to build strategic management measures for the maintenance of vehicle parts so as to get the MTBF (mean time between failures) which is anticipated to provide vehicle maintenance and management personnel with the appropriate timing of maintenance and to help establish the foundation supporting decision-making for the predicted replacement of parts.

2.2 Maintenance Types

In recent years, issues about the maintenance have been widely discussed from various fields. An appropriate maintenance has been playing an important role in work performance and cost-effectiveness of the organization, because under the conditions of increasing the reliability of equipment, to ensure normal manufacturing system processes will help enterprise's overall operating performance [31, 34]. Therefore, proper and timely maintenance is often regarded as key conditions for enterprises to keep their competitive advantage, and different forms of maintenance have become key factors affecting the effectiveness of the maintenance of organization. In general, there are four types of maintenance: corrective maintenance, preventive maintenance, predictive maintenance and proactive maintenance [12]. In the past, companies often used preventive maintenance to reduce the costs of inspection and maintenance, but if not taking into account the current state of the wear and tear of equipment components, this maintenance could easily lead to burden and waste associated with company's business costs [34],

Bousdekis et al. [9] also mentioned that the implementation of preventive maintenance could result in approximately 60% of equipment which have been replaced too early, which may lead to the increase on business costs. Therefore, with the development of industry along with big data, networking and the innovation of information communication technology, the maintenance has been gradually adjusted to predictive maintenance. And with the effective predictive maintenance strategy, the results can be used as the basis for the planning of preventive maintenance period. Thus, this maintenance has been widely promoted and applied in recent years [28]. To sum up, this study focused mainly on preventive maintenance to spot out the risk of failure through model and analysis, and then found out a maintenance period and manner more cost effective and strengthened and to improve the maintenance strategy to achieve business goals with limited maintenance resources.

3 Research Method

3.1 Conceptual Framework

Based on the foregoing content and related literature, this study conducts research on the optimal maintenance replacement cycle for logistics vehicles. First, sort and sort data and clean up the cargo vehicle properties and component maintenance history of the case company, analyze and compare the differences before and after, and select appropriate data quality data to meet the application and scope. Then carry out narrative statistics and chi-square test, and verify whether there is a significant difference in maintenance costs for vehicles of the same property under different circumstances. Subsequently, a model was established using SPSS software, and a decision support model based on a decision tree was established to optimize the replacement cycle. Finally, put forward conclusions and recommendations based on empirical results. See the conceptual framework in Fig. 1.

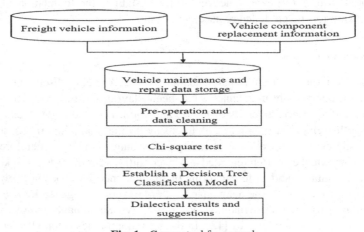

Fig. 1. Conceptual framework

3.2 Research Object

The case company in this research has been established for over 60 years and its main business is to provide Taiwan (including Islands) with cargo transportation services, which is currently one of the top five companies in the logistics industry in Taiwan. At the aspect of the business, this case company has set up satellite operating stations in various regions in order to facilitate the service areas of B2B and B2C. At the aspect of freight delivery, the categories of which include frozen products, frozen goods, food, household appliances, books, daily necessities, clothing and other diversification of commodities. This company currently used the 3.5-ton vehicle as main operating mode in B2C business model; therefore, it is much longer time when the vehicles are assigned and used are mostly need to drive between different factories and regions. Therefore, not just daily routes for vehicles are required to adjusted and planned, but also the degree of wear and tear of the vehicle parts and their replacement are much more frequent. Based on all above, this research aimed a total of 750 3.5-ton level freight vehicles of the case company's logistics services from 2008 to 2016 (total nine years), A 3.5-ton freight car used by the case company's logistics service, with a total of 750 cars (divided into four brands, represented by CT1, CT2, CT3, and CT4) was used as the research and analysis object. The original vehicle resume and maintenance information totaled 190,053. However, in order to seek data quality, quality and efficiency, the maintenance and maintenance details were processed, converted, classified, and normalized by systematic and manual methods. Analysis of quality data. Then, the decision of the optimal replacement cycle of vehicle components is formulated to obtain the appropriate maintenance timing. According to the actual business types, the factory areas of the case company were classified into five types: the coastal area, the industrial area, the metropolitan area, the agricultural area and the comprehensive area. Besides, taking the different geographical environment of the cases into accounts, vehicle parts would be worn to different degree, so this study classified the areas of business stations of the case company into the north, the central, the southern, the eastern and outlying islands areas according to the classification used by the regional meteorological observations and prediction.

4 Data Analysis and Validation

4.1 Data Cleaning and Quality Analysis

This research is based on 3.5 tons of freight vehicles used in the urban logistics in this study. A total of 190,053 vehicle resumes and historical maintenance and repair original data were analyzed. The detailed data sheet records the vehicle registration information actually repaired by each work order. The classified contents 15 types of information such as the area, the factory area, and the name of the business office, the license plate number, the date of issue, the model name, the brand name, the type, year type, tonnage level, current vehicle condition, maintenance date, vehicle components, quantity and amount, ...etc. The content is to be compatible with the data cleaning and subsequent analysis. It is divided into four columns: complete, inconsistent (including format, name, description method, etc.), missing values, and other (including content errors, incorrect fields). After a unified analysis, the data integrity rate is between 71%

and 100% (see Table 1). As shown in the Table 1, inconsistent data (8.99%), missing values (0.02%), and other items (0.03%) account for about 9% of the total data; If the original data was directly brought into the algorithm analysis, it would not possible to establish effective models and classification rules due to incomplete data quality. If the direct deletion method if is used, we will aggregate 32% of parameter (61,428 data) that contains one of following incompleteness: inconsistencies, missing values, and other. Although with the direct deletion method applied and obtained high-quality data (only 68% of parameter), but it also caused large loss of historical data that could result in subsequent analysis and modeling distortion, and furthermore it may result in rules and decisions error.

Table 1. Vehicle maintenance and repair breakdown classification table (original)

Field	Information content (number of entries)				Number of data	Data integrity rate
	complete	Inconsistent	Mission value	others		
Office area	147,580	41,878	251	344		78%
Plant of Business Office	141,367	48,330	117	239		74%
Office name	159,644	30,409	42	97		84%
License number	180,693	9,266	9	85		95%
Date of issue	188,795	1,258	0	0		99%
Vehicle name	190,053	0	0	0		100%
brand name	187,901	2,135	0	17		99%
Type	182,419	7,538	61	35	190,053	96%
Year	149,565	40,488	0	0		79%
Tonnage level	190,053	0	0	0		100%
Car condition	190,053	0	0	0		100%
Maintenance date	169,880	20,173	0	0		89%
Vehicle components	135,280	54,773	0	0		71%
Quantity	190,053	0	0	0		100%
Amount	190,053	0	0	0		100%
Percentage of total	90.97%	8.99%	0.02%	0.03%	100%	

Therefore, in addition to using system analysis, this study attempts to readjust the data generated by the organization in different periods and systems through manual operations; First, filter and sort incomplete vehicle maintenance data, then proceed with data aggregation, conversion, normalization, definitely classify, and define. At the same time, the classification level of large, medium and small items of vehicle components is established in order to obtain better data quality data. The operation rules are summarized as follows:

- In terms of location, pre-adjustment was made based on the final division of business office area, plant area and name.
- Terms of location: pre-adjustment was made based on the final division of business office area, plant area and name.

- Terms of vehicles: the format such as year, type and number is consistent with the content name.
- Terms of maintenance: in addition to the unified date format, in addition to the names of vehicle components, experts recommend using similar and synonymous names for pre-processing of data.
- In connection with the sort of missing value: directly delete to avoid the doubt of data gap.

After repeated experiments in this study, it was found that the classification model constructed was the best in terms of the quality and efficiency of the data collected at the large project level of vehicle components. It took one month to the manual process, and that is 240 man hours, if we count 8 h per day to implement data quality improvement. Therefore, a total of 173,693 maintenance work orders (93.14%) and 23 types of vehicle component items were finally summarized and sorted for the subsequent model construction and system analysis (see Table 2).

Table 2. Vehicle maintenance and repair breakdown table (revised)

Field	Minimum value	Max	average value	Standard deviation	Number of categories	Number of data
Office area	--	--	--	--	5	
Plant of Business Office	--	--	--	--	5	
Office name	--	--	--	--	48	
License number	--	--	--	--	750	
Date of issue	640802	980902	934273	19370	--	
Vehicle name	--	--	--	--	1	
brand name	--	--	--	--	4	
Type	--	--	--	--	12	173,693
Year	2008	2016	--	--	--	
Tonnage level	--	--	--	--	1	
Car condition	--	--	--	--	4	
Maintenance date	970815	1050518	--	--	--	
Vehicle components	--	--	--	--	23	
Quantity	0	851	2	15	--	
Amount	1	159,744	559	2,055	--	

4.2 Analysis of Each Vehicle Brand and the Maintenance Costs

At first, we used chi-squared test to see whether there are any significant differences of the maintenance costs between the vehicles with the same attribute but different brands. According to the five maintenance costs area as shown in Table 3, the costs of each vehicle regardless of its brand have concentrated on the range from $176~$335 and among them, CT3 has taken than the highest proportion(45.5%), CT1(31.0%) the second, CT2(30.7%) the third and CT4 (26.3%) has taken the lowest. The second highest proportion of range of maintenance costs of each brand is respectively from $336~$619

(CT1, 26.0%; CT3, 24.2%), and $ 517 or more (CT2, 23.3%) and from $120~$175 (CT4, 26.0%). According to the chi-squared test, some significant differences in the maintenance costs existed between the vehicles with the same attribute but with different brands.

Table 3. Data analysis statistical table of vehicle brands and their respective maintenance costs

Amount / Brand	$0~$199 (I)		$120~$175 (II)		$176~$335 (III)		$336~$619 (IV)		$620 or more (V)		Total
CT1	12925	12.90%	16655	16.70%	30947	31%	18819	18.80%	20560	20.60%	99906
CT2	9061	12.30%	10974	14.90%	22619	30.70%	17157	23.30%	13839	18.80%	73650
CT3	8	8.10%	7	7.10%	45	45.50%	15	15.20%	24	24.10%	99
CT4	7	18.40%	9	23.70%	10	26.30%	6	15.80%	6	15.80%	38
Total	22001(13%)		27645(16%)		53621(31%)		35997(21%)		34429(20%)		173693

4.3 Each Factory Replacement Vehicle Parts Analysis

The data mining technology was used in this research to have an inquiry in the optimum replacement cycle for vehicle parts in particular factory areas of regions. At first, a statistical data was made from the replacement data of the vehicle parts in each factory, and after the analysis, the top 5 of in each factory area have accounted for the ratio of the overall component for maintenance and replacement more than 68%, in which coastal areas have the highest 93%, and the following are comprehensive areas 88%, agricultural areas 79% industrial areas 74% and metropolitan areas 68%. Among these areas, the item of the most common maintenance in coastal areas is the switch the wiring which accounts for 31% (3,710 times); in metropolitan and comprehensive areas are general maintenance which reaches 39% (23,809 data) and 25% (13,098 data); in industrial and agricultural areas are general repair which reaches 32% (5,800 data) and 27% (4,361 data). The replacement of vehicle parts is as shown in Table 4.

Different areas may lead to differences in the usage of logistics vehicles, so this study has also taken Taiwan's geographical characteristics into consideration (divided into northern, central, southern, eastern and outlying island regions), in order to have an inquiry in the relation of the replacement for vehicle parts between different vehicle brands and different areas. Based on all above, in the study we have found that (as shown in Table 5), in coastal, metropolitan and industrial areas, these two brands CT1 and CT2 in the northern, central and southern regions have accounted for the higher proportion of the maintenance; in agricultural areas, these two brands CT1 and CT2 in the central, southern and eastern regions have accounted for the higher proportion of the maintenance; in the comprehensive areas, these two brands CT1 and CT2 have accounted for the higher proportion of the maintenance in the central and the southern regions. To sum up, although the two brands CT1 and CT2 have always accounted for the highest proportion of the times of vehicle maintenance, there are partial differences when they are in different regions.

Table 4. Table of replacement times of vehicle parts in each factory area (Top 5)

Area \ Item	Coastal Areas	Metropolitan Areas	Industrial Areas	Agricultural Areas	Comprehensive Areas
Vehicle parts	Wiring switch 31%(3,710)	General maintenance 25%(23,809)	General repair 27%(5,800)	General repair 32%(4,361)	General maintenance 39%(13,098)
	General repair 29%(3,421)	General repair 24%(22,757)	Fill light system 25%(4,610)	General maintenance 25%(4,087)	General repair 19%(6,434)
	General maintenance 18%(2,079)	Air conditioning repair 7%(6,694)	Intake and exhaust system 8%(1,491)	Three-stage disassembly 15%(2,405)	Three-stage disassembly 15%(5,060)
	Automotive battery 12%(1,366)	Fill light system 6%(5,668)	Three-stage disassembly 5%(928)	Compartments maintenance 7%(1,152)	Air conditioning repair 11%(3,541)
	Sheet metal 4%(481)	Radiator 5%(5,004)	Clutch 4%(804)	Generator 5%(767)	Front steel plate 3%(1,069)
Total	93% 11,057	68% 69,932	74% 13,633	79% 12,772	88% 29,202

*Remark: the percentage is the number of maintenance times of the vehicle component item in the factory area/the total number of maintenance times of vehicle parts in the factory area; the number in ( ) is the number of maintenance times of the vehicle component item.

Table 5. Maintenance statistical analysis table of vehicle brands in different areas

Factory area \ Region Brand	Northern	Central	Southern	Eastern	Outlying Islands
Coastal area CT1	53.26%	58.45%	40.14%	0	30.67%
CT2	46.70%	41.55%	59.86%	0	69.33%
CT3	0.01%	0	0	0	0
CT4	0.03%	0	0	0	0
Metropolitan area CT1	57.78%	74.70%	52.23%	0	0
CT2	42.12%	25.30%	47.77%	0	0
CT3	0.07%	0	0	0	0
CT4	0.03%	0	0	0	0
Industrial area CT1	58.73%	53.54%	57.54%	0	0
CT2	41.27%	46.46%	42.29%	0	0
CT3	0	0	0.17%	0	0
CT4	0	0	0	0	0
Agricultural area CT1	0	52.12%	75.18%	52.85%	0
CT2	0	47.88%	24.82%	47.15%	0
CT3	0	0	0	0	0
CT4	0	0	0	0	0
Comprehensive area CT1	0	68.20%	70.75%	0	0
CT2	0	31.72%	29.25%	0	0
CT3	0	0.06%	0	0	0
CT4	100.00%	0.02%	0	0	0

4.4 Vehicle Component Classification Model and Performance Assessment Analysis

According to the classification rules for vehicle parts of each factory area analyzed through the model established with the decision tree, 663 rules are listed with the statistical classification (64 in coastal areas, 132 in industrial areas, 127 in metropolitan areas,

150 in agricultural areas and 190 in comprehensive areas). Excerpts from this study that meet the higher probability of occurrence in each field are as follows:

- Rule 1: If the business station is the coastal areas and located in the northern region, the vehicle brand is CT2, the replacement time is longer than 103.5 months but equal to or shorter than 103.6 months, 85.7% of the possibility is to perform general maintenance.
- Rule 2: If the business station is the industrial area and located in the southern region, the vehicle brand is CT2, the replacement time is longer than 108.6 months but equal to or shorter than 109.0 months, 85.7% of the possibility is to perform general maintenance.
- Rule 3: If the business station is the metropolitan area and located in the southern region, the vehicle brand is CT1, the replacement time is longer than 107.7 months but equal to or shorter than 107.8 months, 87.5% of the possibility is to repair air conditioning system.
- Rule 4: If the business station is the agricultural area and located in the central region, the vehicle brand is CT1, the replacement time is longer than 85.3 months but equal to or shorter than 85.4 months, 92.9% of the possibility is to perform 3-stage disassembly.
- Rule 5: If the business station is the comprehensive area and located in the central region, the vehicle brand is CT1, the replacement time is longer than 136.2 months but equal to or shorter than 136.3 months, 90.9% of the possibility is to perform 3-stage disassembly.

5   Discussion

Vehicle parts in this study the construction of predictive maintenance decision model, through the Department of digging a lot of maintenance records, And after data cleaning, to obtain data of appropriate data quality, after research and analysis, the results of maintenance costs for different brands, chi-square inspection, component classification rules, and test data set classification performance were obtained, itemized description as follows:

Terms of Data Quality and Data Scrubbing: Good data quality must possess characteristics such as consistency, correctness, and completeness [2]. Data scrubbing is to remove dirty data and improve its data quality. Although the industry has many tools for data extraction, transformation, and installation, but lacks industry expertise and scalability applications [18]. Based on this, this study found that if didn't scrubbing dirty data from the original data as follow: inconsistent data, missing values, and other, it will not be applicate efficiently. And if the direct deletion method is adopted, the amount of data will be reduced by 61,428, which may lead to distortion of model construction and wrong analysis and decision. Therefore, in this study, a systematic and manual job analysis method was adopted, and also implements relevant expert expertise to data cleaning and sorting operations, which took approximately 240 man hours to process, and after integration, 173,693 suitable data quality data (data totals) were obtained. (Amount 93.14%). On the whole, in the modern era of rapid development of artificial intelligence, the use of workers' wisdom in research methods highlights the fact that enterprises are prolific in organizing data in different systems and periods, coupled with different professional knowledge in various industries. Effective use, most small and medium-sized enterprises, even large organizations, can only adopt the original operation mode, plus the current calculation, analysis, and operation mode, and carry out analysis and research in order to obtain core wisdom and give full play to data. Therefore, in the future, the industry can partially update the system and perform professional manpower operations, which will effectively use the data quality data obtained, and save labor costs and obtain data benefits.

The Vehicle Brands and the Maintenance Costs: Chernatony and McWilliam et al. [11] considered that the brand is consistent with the commitment to quality and also serves as a support for making a purchase decision. Owning the vehicles with reliable performance will be able to save considerable maintenance costs for the enterprise. In this study, we found that there is a significant difference between maintenance costs of the vehicles with the same attribute but not the same brands, On the whole, the maintenance costs of freight logistics vehicles mainly are concentrated on these three ranges $176–$335 (31%), 336–$619 (21%) and $620 or more (20%), which have accounted for 72% of the proportion of overall maintenance times, which shows that the high maintenance costs account for the most proportion of the overall maintenance costs in the case company so there is still more for improvement and review. As far as the brands are concerned, the main maintenance costs of CT1 and CT3 are mostly in the third and the fourth ranges, accounting for respectively 51.6% and 69% of the overall maintenance times, which shows that the maintenance costs of these two brands in the company belong to the medium or higher price; CT2 are mostly in the third and the fifth ranges, accounting for 54% of the overall maintenance times, which shows that the maintenance cost of this brand in the company belongs to the high price; CT4 are mostly in the second and the third ranges, accounting for 50% of the overall maintenance times, which shows that the maintenance cost of this brand in the company belongs to the medium or lower price. On the whole, the comprehensive analysis above has showed that there is a significant difference of the amount of maintenance between the vehicle brands of the case company and the sequence of the maintenance costs of the vehicle brands from high to low is respectively CT2, CT1, CT3 and CT4. Based on this, the organization can review the number, the service life and the dispatch frequency and other performance in the future through the costs and the frequency of the maintenance so as to achieving the purpose of saving maintenance costs and increasing the efficiency of transportation.

The Replacement of Vehicle Parts in Each Factory Area: Generally speaking, in order to prevent the wear, deformation or damage of vehicle parts from leading to problems and concerns the service, which may result in issues about safety and customer complaints and other issues, the general preventive maintenance, repair and other operating items are often scheduled. In this study, we have found that due to the complete government regulations and investment and construction planning, the industrial areas (Industrial parks) have the better and flatter regional roads and customers are more concentrated and thereby the general preventive maintenance and repair and other needs for replacing component are relatively lower (general repair accounted for only 27%); the general maintenance and the general repair and other maintenance items in other areas are listed in the replacement of main components accounting for more than 45% (47% in the coastal areas, 49% in the metropolitan areas, 57% in the agricultural areas and 58% in the comprehensive areas), which has revealed that in order to save the replacement material effectively and to reduce the maintenance costs, the company should take the general maintenance and repair aa the main contents for review and adjustment. And in the coastal areas, because the vehicles were running on the coastal areas with relatively high salinity and humidity, in addition to the general maintenance and repair, switch wiring (31%) has accounted for more than 30% of the replacement needs for components, which has revealed that in the coastal areas, the company should pay more attention to the maintenance against the rust damage and influences on the vehicles due to the salinity and the humidity. And about the maintenance times of vehicles, we have found out that in each factory and area, if taking the brands into consideration, CT1 and CT2 vehicle had higher proportion of maintenance and repair, in which especially the maintenance data of the freight vehicles CT1 in the central metropolitan areas (74.70%), in the southern agricultural areas (75.18%) and in the southern comprehensive areas (70.75%) and other regions were higher than 70%, which has revealed that the maintenance of CT1 was mainly concentrated in the central and southern regions, but which of CT4 was primarily concentrated in the northern comprehensive areas.

Vehicle Parts Classification Model: In this study, we followed the decision tree to analyze and took the areas, factory areas, brands, items, possibility, cycle and other key elements about replacement of vehicle parts into consideration so as to construct an optimal decision support model. In thus study, we have found that the often damages of freight logistics vehicle parts in specific factory areas and regions are related to each other. According to the predictive replacement model, we can obtain the rules and probability of maintenance and replacement in different factory area and different regions. Based on the Decision Support Model, we have constructed a total of 663 classification rules combined with the aforementioned research analysis of vehicle brands, maintenance costs and replacement of vehicle parts to propose the main rules for maintenance and repair. The brands CT1 and CT2 were found to be majority; with the classification in each factory area and region and in comparison with replacement times of vehicle parts, we have found out that the replacement probability of components is higher than 60%, partially more than 90%, which has showed the classification rules through the decision-making model can effectively improve the replacement benefits of vehicle parts.

In summary, in the past, only a few researches were made about predictive vehicle parts replacement decision support for the freight logistics vehicle parts. This study has particularly focused on the maintenance data for all 3.5 ton level freight vehicles to discuss about the rules for maintenance and replacement of vehicle parts then to conduct the data analysis. At the same time, according to the study and analysis of the results of the classification rules, we have verified that the vehicle parts of some particular brands in some specific regions and factory areas were often damaged and in bad condition. The decision support model developed will help the company to establish a reference for the company to create a decision support model to build a predictive optimal replacement cycle. In practice it will help the case company to acquire the appropriate maintenance timing for vehicle parts and the appropriate maintenance frequency as well, and thereby to reduce the maintenance costs and to solve the problem about library reserve stock of vehicle parts, in order to ensure the best condition of operating vehicles and to improve their management and maintenance; on the other hand, it can help the company to determine if the company should continue to maintain these vehicles or scrap the old vehicles or purchase new vehicles, etc., and can clarify the condition that specific vehicle parts in the business station of the case company are often damaged in different factory areas and different regions. On the whole, the study results can help vehicle managers and technicians to obtain maintenance experience; moreover, with the knowledge sharing and storage, the company will be able to solve the problems about uneven quality of maintenance and about technicians' insufficient maintenance ability. With the decision support for an optimum replacement cycle of vehicle parts, the organization can save the costs related to maintenance.

6 Conclusion and Suggestions

With the growing of economy, the logistics industry has been committed to achieving the "Last mile" like physical transportation and distribution services, which is the important connotation of supply chain activities [30]. Therefore, the competitiveness among the logistics industries has become very fierce, so the poor efficiency of goods distribution will represent resources wastes and the weak competitiveness. Only a few researches were made about predictive vehicle parts replacement decision support for the freight logistics vehicle parts. The case company in this study accounted for one of top five proportion of Taiwanese freight logistics market share. Use its complete vehicle maintenance information to obtain good data quality data through system and manual data cleaning methods, And combined with data mining classification methods for analysis, and then according to the classification rules and the analysis results to verify that the vehicle parts in some specific regions and factory areas were often damaged and in bad condition when being maintained or replaced. The decision support model developed in this study will help the company to establish a reference for the company to create a decision support model to build a predictive optimal replacement cycle. This will help the organization to acquire the appropriate maintenance timing for vehicle parts and to confirm the appropriate maintenance frequency, it has also taken the corporate business model into account to provide vehicle maintenance technicians and management personnel maintenance knowledge and ability which are different from the previous standard of care standards or don't only rely on rules of thumb. Therefore, through this study, the company can reduce the maintenance costs and to solve the problem about library reserve stock of vehicle parts, in order to ensure the best condition of operating vehicles and to improve their management and maintenance. On the other hand, it can help the company to determine if the company should continue to maintain these vehicles or scrap the old vehicles or purchase new vehicles or other decisions.

According to the study, we have found out that there are some significant differences of maintenance costs between the vehicles of the case company with the same attribute but different brands; the vehicle parts of some particular brands in some specific regions and factory areas were often damaged and in bad condition. The decision support model developed in this study will help the company to establish a reference for the company to create a decision support model to build a predictive optimal replacement cycle. On the whole, the study results can help supervisors and technicians to obtain maintenance experience; moreover, with the knowledge sharing and storage, the company will be able to solve the problems about uneven quality of maintenance and about technicians' insufficient maintenance ability. Meanwhile, with the decision support for the optimum replacement cycle of vehicle parts. In addition, through the data quality cleanup operation, we can learn that the company's data can be used to complete the uniform, standardized and unified use of the system platform interface, in addition to saving labor and operation costs, and can effectively use high-performance data quality management methods, so that the data generated within the organization can be fully utilized to achieve comprehensive results, and then achieve business optimization, cost energy efficiency and utility optimization of corporate goals, so as to achieve enhanced competitive advantage To ensure the goal of sustainable operation. In summary, in the future we can develop a more appropriate predictive replacement model of the vehicle parts combined with the model of freight vehicles with different types and tonnages level, the drivers' driving habits, the vehicle running state and other information.

References

1. Agresti, A., Kateri, M.: Categorical data analysis. In: Lovric, M. (ed.) International Encyclopedia of Statistical Science, pp. 206–208. Springer, Heidelberg (2011). https://doi.org/10.1007/978-3-642-04898-2_161
2. Aebi, D., Perrochon, L.: Towards improving data quality. In: Sarda, N.L. (ed.) Proceedings of the International Conference on Information Systems and Management of Data, Delhi, pp. 273–281 (1993)
3. Berry, M.J., Linoff, G.: Data Mining Technique: For Marketing, Sales, and Customer Relationship Management. Wily Press, New York (1997)
4. Bose, I., Mahapatra, R.K.: Business data mining-a machine learning perspective. Inf. Manag. **39**(3), 211–225 (2001)
5. Bousdekis, A., Papageorgiou, N., Magoutas, B., Apostolou, D., Mentzas, G.: Enabling condition-based maintenance decisions with proactive event-driven computing. Comput. Ind. **100**, 173–183 (2018)
6. Brachman, R.J., Khabaza, T., Kloesgen, W., Piatetsky-Shapuro, G., Simoudis, E.: Mining business database. Commun. ACM **39**(11), 42–48 (1996)
7. Chen, M.C., Hsu, C.L., Chang, K.C., Chou, M.C.: Applying Kansei engineering to design logistics services–a case of home delivery service. Int. J. Ind. Ergon. **48**, 46–59 (2015)
8. Chernatony, L.D., McWilliam, G.: Branding terminology the real debate. Mark. Intell. Plan. **7**(7), 29–32 (1989)
9. De Faria, H., Costa, J.G.S., Olivas, J.L.M.: A review of monitoring methods for predictive maintenance of electric power transformers based on dissolved gas analysis. Renew. Sustain. Energy Rev. **46**, 201–209 (2015)
10. Farid, D.M., Zhang, L., Rahman, C.M., Hossain, M.A., Strachan, R.: Hybrid decision tree and naïve Bayes classifiers for multi-class classification tasks. Expert Syst. Appl. **41**(4), 1937–1946 (2014)
11. Fayyad, U., Poatersky-Shapiro, G., Symth, P.: From data mining to knowledge discovery in databases. AI Mag. **17**, 37–54 (1996)
12. Ghajargar, M., Zenezini, G., Montanaro, T.: Home delivery services: innovations and emerging needs. IFAC-PapersOnLine **49**(12), 1371–1376 (2016)

13. Han, J., Pei, J., Kamber, M.: Data mining: concepts and techniques. Elsevier, Amsterdam (2011)
14. Galhardas, H., Florescu, D., Shasha, D., Simon, E., Saita, C.A.: Declarative data cleaning: language model and algorithms. In: Proceedings of the 27th International Conference on Very Large Databases (VLDB 2001), pp. 371–380 (2001)
15. Jim, W.: The butterfly effect of data quality. In: The Fifth MIT Information Quality Industry Symposium (2011)
16. Lee, X., Massman, W., Law, B.: Handbook of Micrometeorology. A Guide for Surface Lux Measurement and Analysis. Springer, Dordrecht (2005). https://doi.org/10.1007/1-4020-2265-4
17. Cai, L., Zhu, Y.: The challenges of data quality and data quality assessment in the big data era. Data Sci. J. **14**(2), 1–10 (2015)
18. Ohbyung, K., Namyeon, L., Bongsik, S.: Data quality management, data usage experience and acquisition intention of big data analytics (2014)
19. Schmid, H.: Probabilistic part-ofispeech tagging using decision trees. In: New Methods in Language Processing, p. 154. Routledge (2013)
20. Shaw, M.J., Subramaniam, C., Tan, G.W., Welge, M.E.: Knowledge management and data mining for marking. Decis. Support Syst. **31**(1), 127–137 (2001)
21. Shin, J.H., Jun, H.B.: On condition based maintenance policy. J. Comput. Des. Eng. **2**(2), 119–127 (2015)
22. Taleb, I., Dssouli, R., Mohamed, A.S.: Big data pre- processing: a quality framework. In: 2015 IEEE International Congress on, pp. 191–198 (2015)
23. Tilahun, N., Thakuriah, P.V., Li, M., Keita, Y.: Transit use and the work commute: analyzing the role of last mile issues. J. Transp. Geogr. **54**, 359–368 (2016)
24. Waeyenbergh, G., Pintelon, L.: A framework for maintenance concept development. Int. J. Prod. Econ. **77**(3), 299–313 (2002)
25. Wang, Y.F., Zhang, C.Z., Zhang, B.B., et al.: A survey of data cleaning. New Technol. Libr. Inform. Serv. **12**, 50–56 (2007)
26. Wu, I.C., Li, R.J., Chen, T.L.: A vehicular maintenance and replacement decision support system in distribution services: a data mining technique. J. Manag. Syst. **21**(1), 111–137 (2014)
27. Zheng, Z., Zhou, W., Zheng, Y., Wu, Y.: Optimal maintenance policy for a system with preventive repair and two types of failures. Comput. Ind. Eng. **98**, 102–112 (2016)

An Efficient and Metadata-Aware Big Data Storage Architecture

Rize Jin[1($\boxtimes$)], Joon-Young Paik[1], and Yenewondim Biadgie[2]

[1] School of Computer Science and Technology, Tiangong University, Tianjin 300160, China
jinrize@tiangong.edu.cn
[2] Department of Software and Computer Engineering, Ajou University,
Suwon 16499, Republic of Korea

Abstract. This paper introduces a hash partitioning-based file compaction design to improve the efficiency of storing and accessing small files in big data storage systems. The proposed approach consists of a file compaction tool and an access interface. The compaction tool merges a group (usually a directory) of small files into a set of "big files" to reduce the metadata required to be maintained in the on-chip memory. The data locality and tree structure of those small files are preserved. The access interface is designed to provide transparent access to the small files in the big files. Experimental results confirm that the proposed approach lead to a significantly enhancement in terms of namespace usage and access speed.

Keywords: Big data · Small file · File compaction · Metadata management

1 Introduction

A file is called small [1] when its size is substantially less than the block size of big data platforms [2, 3] which typically is 128 KB. Files and blocks are metadata objects and occupy namespace (mapping space) of the on-chip memory. When there are numerous small files stored in the system, the massive metadata information can occupy a large memory space. The storage space may be underutilized because of the namespace limitation. For example, 100 million files with an average size of 3 K can consume 70% of the namespace capacity of 256 MB RAM, yet occupy only 1% of a 4 TB storage. Moreover, massive small files generate small and random writes, which incurs an increase of write amplification [4]. This paper proposes a file compaction method as an effective solution to the problem of managing massive amounts of small files in big data storage systems. The proposed approach increases the scalability of the system by reducing the namespaces usage and decreasing the operation load in the on-chip memory by distributing namespace management.

Our primary contribution is to explore empirically a hash partitioning-based file compaction format to organize a group of small files into a compact file consisting of multiple larger files. The file compaction reduces the amount of file metadata residing in RAM. The compact file retains data locality and data intact by maintaining a local index and the metadata information of the original small files. Our second contribution

Y. Nah et al. (Eds.): DASFAA 2020 Workshops, LNCS 12115, pp. 146–152, 2020.
https://doi.org/10.1007/978-3-030-59413-8_12

is to provide a set of principles to implement a multi-thread merge tool to improve the efficiency of creating "big files" from many small files. Each big file is a self-manageable key-value document.

The remainder of this paper is organized as follow. Section 2 defines the problem of managing massive amounts of small files in big data storage. Section 3 presents the design details of the proposed solution. Section 4 reports the performance evaluation. Section 4.1 discusses related work and Sect. 5 concludes this paper.

2 Problem Definition

Several studies [5–7] indicate that small and random writes are slower than sequential writes. Small files tend to incur many small and random writes. We call a write small when the request size is equal to or less than the block size (i.e., <128 KB). The controller performs additional work to maintain the metadata necessary for mapping small writes. Further, an enormous number of small files requires significantly more namespace to maintain the extremely large amount of metadata and this is clearly not efficient as the big data storage systems have a relatively limited on-chip memory space [8, 9]. Read performance is a consequence of the write pattern. When writing a large chunk of data, it is likely to be stored in locations that are contiguous in the physical space. However, small files whose addresses are contiguous in the logical space may refer to addresses that are not contiguous in the underlying storage system to the dynamic mapping performed by the buffer management.

The on-chip memory is relatively a limited resource. The OS maintains namespace information for each file buffered in the on-chip memory. When the number of files becomes extremely large, the usage of the memory increases sharply. For example, assume that there are 100 million small files and the metadata of each file occupies 100 bytes of memory space; then, they would consume 1 GB memory space. As user data increases, it is not difficult to exceed the namespace limit. The actual namespace stores considerably more information including the metadata of directories and blocks.

3 The Proposed Method

3.1 Design Considerations

Considering the abovementioned issues, the proposed file compaction approach includes five design principles: 1) **Buffer hot files**. Files that change frequently are considered as hot [10]. Hot files and their metadata should be buffered in the on-chip memory as much as possible and written to the underlying storage infrequently. To address this, the proposed method uses the least recently used (LRU) buffer replacement policy. 2) **Collect small writes**. To maximize throughput and minimize write amplification, whenever possible, small writes should be maintained in the buffer in the on-chip memory and only written once when the buffer is full. 3) **Clean-first buffer management** [11]. The buffer manager should consider not only the buffer hit ratio but also the heavy write cost of the logging mechanism employed by the majority of big data systems. 4) **In-block update**. It uses an in-block update (IBU) strategy that divides one physical block into data blocks and log blocks to maintain the updates of small files both separately from and locally to the original files. **5) Prefetching** [12]. It is a widely used storage optimization technique to reduce buffer misses by exploiting access patterns and fetching data into a buffer before they are requested. We employ a two-level prefetcher, which consists of local index file prefetching and correlation-based file prefetching.

3.2   Compaction File Format

Figure 1 illustrates a method to locate a small file in the proposed method. The small file is stored in a compact file that consists of multiple big files. Small files within big files are indexed by a hash index to maintain the original separation of data. In detail, small files are stored as a set of <key, value> pairs, the file name (or the path of the file in a host system) is the key and the content is the value. A metadata file records the original directory tree structure.

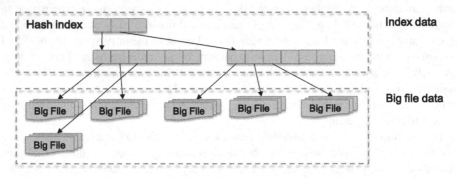

Fig. 1. Locate small files using hash index.

3.3   Transparent Access

The proposed method is designed to fit into the underlying file system and hence, can provide transparent access to the upper applications. When accessing a directory, the user can expose small files and the directory tree structure. Individual small files can also be accessed directly without being extracted from the big file. Accessing a file is slower than a usual access. We first must locate the index of the small file and then read the data from the big file. The application or host uses the absolute path of the small file as the identifier (sID) when accessing. The hash function calculates a bID value, which is the identifier of its corresponding big file, from sID and the system can then locate the local index file by finding a record that associates with bID in the namespace. Figure 2 presents the process.

3.4   File Compaction Tool

To create a big file efficiently, this paper implements a file compaction tool in a multi-thread fashion: a list of small files is generated by traversing the source directories

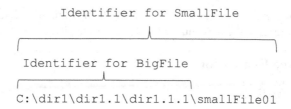

```
                  Identifier for SmallFile

      ┌──────────────────────┴──────────────────────┐
              Identifier for BigFile

      ┌──────────┴──────────┐
      C:\dir1\dir1.1\dir1.1.1\smallFile01
```

Fig. 2. Resolve *sID* to *bID*.

recursively and then the list is divided into several ranges using a hash function. The small files within each range are stored into a big file (approximately 8 MB, configurable). Finally, index and metadata files are generated. In detail, each big file consists of a metadata file and a local index file, which records the offset and length for each original small file it includes, and a set of <key, value> pairs, where key is the file name and value is the content of that file. The reason the index and the metadata file are maintained within the big file is that we want these data stored physically close together to facilitate prefetching.

Compared to maintaining numerous metadata objects as a global index in the namespace, the proposed solution does not result in additional overhead to the on-chip memory. If the sum of the small files to be compacted exceeds the predefined size for a big file, the list of small files is divided into multiple big files. In Fig. 3, a directory containing many small files is combined as a directory with several big files and indexes.

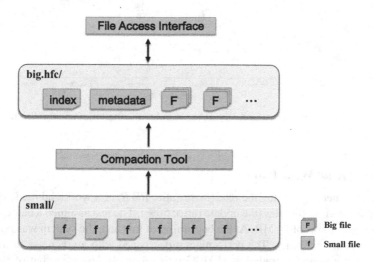

Fig. 3. Compact small files into a big file.

As mentioned earlier, the big data storage functions well for a sequential access pattern. However, small files usually do not provide sequential accessing, even for a batch processing requested to them. The reason is that they are likely to lose data locality in the presence of out-of-place updates. Consequently, the host cannot provide a prefetching function. File compaction ensures that related small files are stored sequentially and in-block updates preserve the maximum possible data locality.

4   Performance Evaluation

4.1   Experiment Settings

The test platform was built on a Core i5–2500 machine (3.30 GHz, 16 GB DDR3 RAM) running Ubuntu 16.04 with Hadoop version 2.7.

Figure 4 compares the memory usage of the proposed method (HFC) and the original approach when the system stored multiple file sets. We varied the number of small files, *#ofSmallFiles*, along with the sizes of the big files, *sizeOfBigFile*. As expected, the proposed approach achieved considerably improved efficiency storing small files compared to the original method, increasing approximately 510 times for *sizeOfBigFile* = 1.75 MB and approximately 3,800 times for *sizeOfBigFile* = 14 MB. The proposed approach required less namespace because of its local index file design.

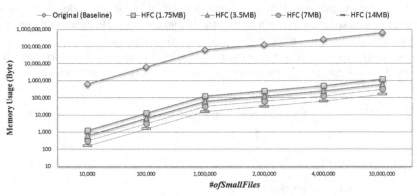

Fig. 4. Comparison of memory usage varying the number of small files and the size of big files.

4.2   Varying Read/Write Ratio

HFC was evaluated by varying the read/write ratio, *RWR*, of a workload made of a mix of interleaved reads and writes (the update ratio: 50%). The test assumed a namespace of 32 MB and buffer size of 64 MB. At the beginning of this test, the system was populated with two million small files. The performance decreased with an increase in the write ratio. The performance degradation of HFC was relatively less than that of the other approach, as indicated in Fig. 5. The reason is that the in-block update of HFC reserves the data locality of the original file and its logs and therefore, HFC performs fewer disk writes. Conversely, the existing approach can suffer performance degradation owing to the increased garbage collection ratio. In particular, the execution time of the original approach increased 100 times as the *RWR* increased from 100/0 to 0/100; HFC had an increase of 36%.

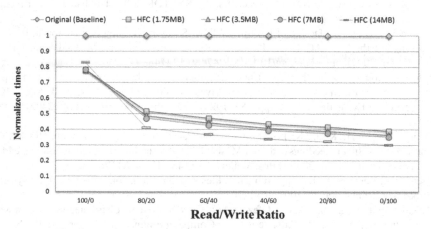

Fig. 5. Comparison of execution times by varying read/write ratio.

5 Conclusions

The increasing capacity of big data storage systems has resulted in significant over-head for managing the in-RAM metadata of massive amounts of small files. To ensure the scalability and efficiency of accessing these small files, this paper proposed a hash partitioning-based file compaction: a novel file compaction interface that provides transparent access to individual small files within a large file using a local index design. In detail, massive amounts of small files are compacted into a small number of big files; each big file consists of a local index file that records the offset and length for each original small file included and a set of <key, value> pairs, where key is the file name and value is the content of that file. Compared to maintaining many metadata objects as a global index in the namespace, the proposed solution reduces significantly the lookup and access overhead to big data storage systems.

Acknowledgments. This work was supported in part by the National Natural Science Foundation of China (NSFC) under Grant 61806142, in part by the Natural Science Foundation of Tianjin under Grant 18JCYBJC44000.

References

1. Jouppi, N.P.: Improving direct-mapped cache performance by the addition of a small fully-associative cache and prefetch buffers. ACM SIGARCH Comput. Architect. News **18**, 364–373 (1990)
2. McAfee, A., Brynjolfsson, E., Davenport, T.H., Patil, D.J., Barton, D.: Big data: the management revolution. Harvard Bus. Rev. **90**(10), 60–68 (2012)
3. Labrinidis, A., Jagadish, H.V.: Challenges and opportunities with big data. Proc. VLDB Endowment **5**(12), 2032–2033 (2012)

4. Hu, X.Y., Eleftheriou, E., Haas, R., Iliadis, I., Pletka, R.: Write amplification analysis in flash-based solid state drives. In: Proceedings of SYSTOR 2009: The Israeli Experimental Systems Conference, p. 10 (2009)
5. Lee, S.W., Moon, B.: Design of flash-based DBMS: an in-page logging approach. In: Proceedings of the 2007 ACM SIGMOD International Conference on Management of Data, pp. 55–66 (2007). https://doi.org/10.1145/1247480.1247488
6. Chen, F., Koufaty, D.A., Zhang, X.: Understanding intrinsic characteristics and system implications of flash memory based solid state drives. In: Proceedings of ACM SIGMETRICS Performance Evaluation Review, pp. 181–192 (2009). https://doi.org/10.1145/2492101.155 5371
7. Kang, J.U., Jo, H., Kim, J.S., Lee, J.: A superblock-based flash translation layer for NAND flash memory. In: Proceedings of the 6th ACM & IEEE International Conference on Embedded Software, pp. 161–170 (2006). https://doi.org/10.1145/1176887.1176911
8. Bende, S., Shedge, R.: Dealing with small files problem in hadoop distributed file system. Procedia Comput. Sci. **79**, 1001–1012 (2016)
9. ElKafrawy, P.M., Sauber, A.M., Hafez, M.M.: HDFSX: big data distributed file system with small files support. In: 2016 12th International Computer Engineering Conference (ICENCO), pp. 131–135. IEEE, 28 December 2016
10. Park, D., Du, D.H.: Hot data identification for flash-based storage systems using multiple bloom filters. In: 2011 IEEE 27th Symposium on Proceedings of Mass Storage Systems and Technologies (MSST), pp. 1–11 (2011). https://doi.org/10.1109/msst.2011.5937216
11. Jin, R., Cho, H.J., Chung, T.S.: LS-LRU: a lazy-split LRU buffer replacement policy for flash-based B+-tree index. J. Inf. Sci. Eng. **31**(3), 1113–1132 (2015)
12. Joseph, D., Grunwald, D.: Prefetching using Markov predictors. In: Proceedings of ACM SIGARCH Computer Architecture News, pp. 252–263 (1997). https://doi.org/10.1109/12.752653

Kernel Design of Intelligent Historical Database for Multi-objective Combustion Optimization

Wei Zheng[1,2] and Chao Wang[1(✉)]

[1] Tianjin Key Laboratory of Process Measurement and Control,
School of Electrical and Information Engineering, Tianjin University, Tianjin 300072, China
zhengwei@th.tjtc.edu.cn, wangchao@tju.edu.cn
[2] School of Mechatronical Engineering and Automation,
Tianjin Vocational Institute, Tianjin 300410, China

Abstract. In order to make the methods and strategies of artificial intelligence with data-driven and massive historical operation data combine directly with each other, and then meet the intelligent demand of multi-objective combustion optimization for coal-fired power stations, an intelligent historical database (IHDB), which is a combination of artificial intelligence and historical database, is proposed. The structures of file, index and module, which compose the kernel of IHDB, are presented as well. The kernel design doesn't only ensure processing data efficiently, but also incorporates data analysis, optimization decisions and other intelligent elements, so that a strong platform can be developed for solving the problem of multi-objective combustion optimization. The performance tests about storing and querying data of IHDB have obtained satisfactory results. Developing IHDB is beneficial to improve the intelligent level of coal-fired power plants and is helpful to promote the development of real-time database system involving IHDB.

Keywords: Historical data base · Data-driven · Artificial intelligence · Multi-objective combustion optimization · Kernel structure

1 Introduction

With the wide application of distributed control system (DCS) and real-time database system (RTDBS), the levels of automation and information of coal-fired power plants have been significantly improved. The main problem has become system optimization from automatic control and data integration. Combustion optimization, which belongs to system optimization, can be defined to a multi-objective optimization problem for reducing pollutant emissions and increasing boiler efficiency. At present, multi-objective combustion optimization based on data-driven has been widely studied, and significant progress has been made in data mining [1–6], data modeling [7–12], evolutionary computing [13–15] and data-driven hybrid strategy [16]. However, the above studies only select some historical operation data from the historical database (HDB) for the research work, so that there is a defect in data integrity. Moreover, the research results are lack

© Springer Nature Switzerland AG 2020
Y. Nah et al. (Eds.): DASFAA 2020 Workshops, LNCS 12115, pp. 153–163, 2020.
https://doi.org/10.1007/978-3-030-59413-8_13

of practical application. In addition, HDB generally doesn't own the abilities of data mining, data modeling and evolutionary computing. In that case, the research results of multi-objective combustion optimization based on data-driven can't be closely combined with HDB, so that the massive historical operation data have not been effectively utilized, either.

Therefore, it is necessary to integrate the methods and strategies of artificial intelligence into HDB in order to make HDB clever enough to automatically finish mining, modeling and optimizing, in order that an intelligent historical database (IHDB) can be produced by combining artificial intelligence with historical database. Actually, *Chen* et al. developed the KDPAG intelligent database which had combined knowledge base, data base, pattern recognition, artificial neural network and genetic algorithm together in 1997 [17]. Later, the KDPAG intelligent database was applied on alumina production successfully [18]. *Abel* et al. established an intelligent database by combining data base system with knowledge base system to analyze and manage petrographic data [19]. *Nihalani* et al. discussed the approaches for developing the intelligent database with artificial intelligence and data base, and investigated the progress in the field of intelligent database [20]. The above studies provide a strong reference for exploiting IHDB. In addition, with the development of industrial big data, smart plant and intelligent manufacturing, only when coal-fired power plants and other process factories effectively process massive historical operation data with all kinds of intelligent methods, can they keep up with the era [21–23].

Above all, the intelligent historical database (IHDB), which is a combination of artificial intelligence and historical database, is proposed in this paper. Based on the idea of IHDB, the main aim of this work is to present the kernel design of IHDB, which can't only efficiently cascade the basic functions of storing, querying, analyzing etc., but also is able to integrate data mining, data modeling and evolutionary computing into the IHDB. Therefore, IHDB can meet the intelligent demands by itself, such as multi-objective combustion optimization.

2   Design of Kernel Structure

It is crucial for IHDB to have a stable and efficient kernel structure, which can improve the efficiency of processing massive historical data and easily combine intelligent computation. Moreover, the kernel structure of IHDB (namely HDB) is related to the overall performance of RTDBS [24]. The detailed discussion will focus on the design of file structure, index structure and module structure of IHDB kernel as following.

2.1   Design of File Structure

As we know, historical data are stored in the disk files. The IHDB proposed in this paper consists of two kinds of disk files, which are history data file and management information file. IHDB contains many history data files, which mainly store the historical data of different tags and the index information of these historical data. Where a tag in the IHDB is corresponding to an equipment in the plant. There is only one management information file in the IHDB, which is used to save the information of all the history data files and the index among them. Therefore, the relationship between management information file and history data file is one-to-many as shown in Fig. 1. According to the query time, IHDB can locate the corresponding historical data file by management information file.

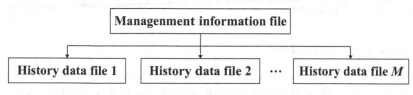

Fig. 1. File structure of IHDB

A history data file records the historical data of all the tags within a time period. That's to say, the historical data of different tags within the same time period are only stored in one history data file. The history data file manages data with page storage in order to improve the efficiency of storing and querying. Where the size of a page is usually set to an integer multiple of 1K, and it is set to 4K in this paper. A history data file, whose structure is demonstrated as Fig. 2, consists of three parts: file beginning part, tag index part and data information part.

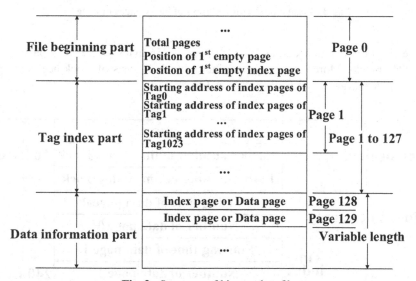

Fig. 2. Structure of history data file

In Fig. 2, file beginning part is the starting page of a history data file, which records many information about the history data file, including total pages, the position of first empty page, the position of first empty index page and so on. Tag index part is a part with the fixed length of 127 pages and preserves the starting address of index pages of every tag as an array. Data information part is a part with a variable length and consists of a mass of index pages and data pages. Where index page is used to establish a sequential time index for historical data and it is composed of many time index blocks. Moreover, every time index block contains several time indexes and a time index is corresponding to a data page in the history data file. Data page stores historical data and the historical data of a tag can be kept in different data pages, however, the historical data of different tags can't be stored in the same data page.

2.2 Design of Index Structure

In this paper, index structure adopts an index mechanism with three levels based on time, the first level is to retrieve from management information file to a history data file; the second one is to search from tag index part to an index page; the last one is to locate the data page by its time index.

The starting address of index pages of each tag in tag index part takes up 4 bytes and its format is displayed as Fig. 3. Since the size of every page is 4K, every page in tag index part has 1024 starting addresses of index pages of different tags (1024 tags).

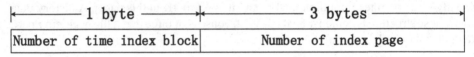

Fig. 3. Format of starting address of index pages of a tag

The size of a time index block in index page is 256 bytes, so an index page has 16 time index blocks. Moreover, every time index block consists of block beginning part and time index array, which is illustrated as Fig. 4.

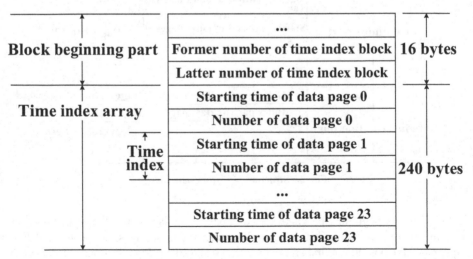

Fig. 4. Structure of time index block

Every time index block has 24 time indexes. Each time index consists of the start time of data page and the number of data page and takes up 10 bytes. Therefore, data page, which stores the desired data, can be located by its corresponding time index. Above all, the retrieval process of index mechanism with three levels is shown as Fig. 5.

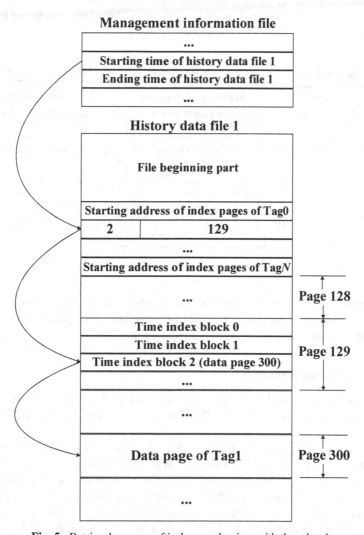

Fig. 5. Retrieval process of index mechanism with three levels

Figure 5 is an example to query the historical data of Tag1 within a certain time period. First, the tag index part of history data file 1 is located by management information file. Next, the time index block can be found in the index page of data information part according to the starting address of index pages of Tag1. Finally, the data page can be determined by the corresponding time index in the time index block. A data page consists of some pieces of historical data and every piece of historical data contains three items, including time, quality and value. After locating the data page, the desired historical data can be obtained by binary searching or sequential searching depend on the querying time.

Additionally, time index blocks stored in the index page belong to one tag, and all the time indexes of one tag are able to generate an index chain, which can make all the data pages of one tag link together, so that it is easy and fast to process all the historical data. The structure of an index chain is shown as Fig. 6.

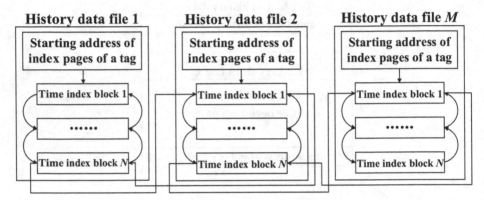

Fig. 6. Structure of index chain

As shown in Fig. 6, the index chain is able to achieve linking from double sides, so that it is convenient to locate the position of historical data, which need to be processed. What's more, the index chain can across the history data files.

2.3 Design of Module Structure

Module structure is adopted to design the kernel of IHDB, so that it is convenient to develop, maintain and improve IHDB. The framework of IHDB kernel is demonstrated as Fig. 7.

In Fig. 7, interface module is to provide an access for external applications to call the various functions of IHDB. Writing module is mainly used to receive the input historical data and reading module is primarily used to output the queried historical data.

Compression module is responsible for compressing analog data and digital data. Analog data are compressed by spinning door transformation (SDT) algorithm, and digital data will be saved only when there is one or more bits changing. Although IHDB doesn't keep all the sample data, it can recover them within the specified accuracy range as other HDBs [25].

Access module helps writing module and reading module transmit historical data by operating the index. Visiting module of management information is used to query and update the content of management information file. The data and information of history data file can be stored and queried by the visiting module of history data.

To avoid operating disk frequently and increase execution speed, buffer-pool module is designed based on the buffer-pool template of Berkeley DB [26] to build a buffer in

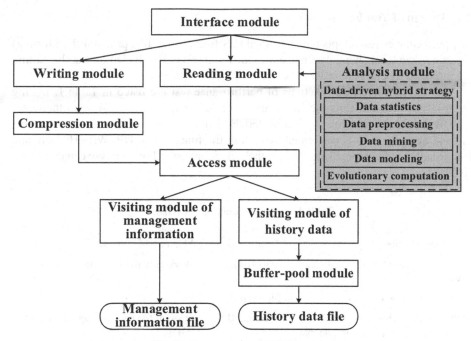

Fig. 7. Framework of IHDB kernel

memory in this paper. If a storage operation is performed, the historical data to be written to disk are placed in the buffer at first. When historical data are accumulated to a certain amount in the buffer, then these historical data will be written one-time to the history data file in the disk. If a query operation is executed, the buffer will be checked for the queried historical data firstly. If not, historical data will be searched from the history data file.

Analysis module can achieving the intelligent computation of IHDB. According to the historical data obtained by reading module, analysis module applies the methods of intelligent computation (including data statistics, data preprocessing, data mining, data modeling, evolutionary computing and so on) on these data on the one hand and implements the data-driven strategy on the other hand. In that case, historical data and intelligent computation can combine with each other closely, which will be effective to drive the artificial intelligence and intelligent manufacturing with data.

3 Performance Test of IHDB Kernel

In order to improve the portability and compatibility, IHDB kernel is coded in C language, and the present kernel of IHDB only takes up 1081 KB by optimizing the code overall. Additionally, the relevant experiments of intelligent computation of IHDB has been conducted in the earlier works, which were demonstrated in reference [16]. The following tests focus on the performance of IHDB kernel, i.e. the performance of storing and querying historical data.

3.1 Design of Test Framework

The performance test adopts the traditional C/S framework. Test program (Exe format) is taken as the local client, which operates the local server IHDB (Dll format) by calling the interface functions in order to conduct the performance test.

The relevant interface functions of performance test are listed in Table 1. During the test, test program calls the DB_StartServer function to start IHDB, and then uses the DB_InitServer function to initialize IHDB. After that, the performance test on storing and querying data is performed by calling the functions of DB_WriteHisData and DB_ReadHisData respectively. At last, the DB_StopServer function is executed to stop IHDB.

Table 1. Interface functions of IHDB

Interface function	Action	Meaning of variable
int32_t DB_StartServer(const char *WorkPath)	Start IHDB	WorkPath: the working path of IHDB
int32_t DB_StopServer()	Stop IHDB	
int32_t DB_InitServer(u_int32_t nCount, u_int32_t *pTagID, double *pCompressInfo)	Initialize IHDB	nCount: the count of tags existed in IHDB *pTagID: the array of IDs, which keep the IDs of tags existed in IHDB pCompressInfo: the array of compression ratios, which keep the compression ratios of tags existed in IHDB
int32_t DB_WriteHisData(u_int32_t nTagID, u_int32_t nCount, InsDataStru *pHisData)	Write data	nTagID: tag ID nCount: the count of tags to be written to IHDB *pHisData: the array of data to be written to IHDB, where InsDataStru is a user-defined structure
int32_t DB_ReadHisData(u_int32_t nTagID, MyTimeDef startTime, int32_t qTime, u_int32_t *nRetCount, InsDataStru *pHisData)	Read data	nTagID: tag ID startTime: querying time, including second and millisecond, MyTimeDef is a user-defined structure qTime: the length of querying time *nRetCount: returns the count of data obtained by querying *pHisData: returns the array of data obtained by querying

3.2 Results of Performance Test

The performance test is conducted under different conditions. Due to space limitations, only one case is given here, whose condition contains operating system (Windows7, 32 bits), CPU (i5 processor), the size of memory (2G), the format of hard disk (NTFS), the size of every time index block (256 bytes), the size of each page (4K), the size of buffer (400000K, i.e. 100000 pages). The results of performance test under the above condition are illustrated in Table 2 and Table 3.

Table 2. Test results of storing performance

Description	Executions	Time (s)	Speed (executions /s)
Test of storing 1 piece of data of one tag	200000000	956.96	208,994
Test of storing 10 pieces of data of one tag	200000000	495.90	403,309
Test of storing 100 pieces of data of one tag	200000000	492.80	405,846
Test of storing 1000 pieces of data of one tag	200000000	504.08	396,764

Table 3. Test results of querying performance

Description	Executions	Time (s)	Speed (executions /s)
Test of querying 1 piece of data of one tag	1000000	7100.74	1,408
Test of querying 10 pieces of data of one tag	1000000	721.30	13,864
Test of querying 100 pieces of data of one tag	1000000	77.46	129,099
Test of querying 1000 pieces of data of one tag	1000000	14.77	677,245

Under the same condition, the comprehensive results of performance test show that the hard disk format of NTFS is a little better than FAT32, the best size of a time index block is 256 bytes, the best size of a page is 4K and 400000K is enough for the size of buffer.

4 Conclusion

Face to intelligent manufacturing, in order to organize and utilize massive historical operation data effectively to solve intelligent problems, such as multi-objective combustion optimization, coal-fired power plants do not only need historical database to provide efficient storing and querying functions, but also to have intelligent abilities including data analysis, optimization decision and so on. Intelligent historical database is proposed in this paper by combining the research results of artificial intelligence with massive data of historical database directly. Based on the idea of IHDB, the structures of file, index and module, which compose the kernel of IHDB, are presented. What's more, the methods and strategies of artificial intelligence including data mining, data modeling, evolutionary etc. are integrated into IHDB. The performance tests about storing and querying data of IHDB have obtained satisfactory results. The IHDB, which is based on kernel structure designed in this paper, can meet a lot of intelligent demands including multi-objective combustion optimization and make data-driven artificial intelligence more practical. Moreover, IHDB is able to further support the real-time database system and provide a reference for other process industries to improve the intelligent level.

References

1. Kusiak, A., Song, Z.: Combustion efficiency optimization and virtual testing: a data-mining approach. IEEE Trans. Ind. Inform. **2**(3), 176–184 (2006)
2. Song, Z., Kusiak, A.: Constraint-based control of boiler efficiency: a data-mining approach. IEEE Trans. Ind. Inform. **3**(1), 73–83 (2007)
3. Kusiak, A., Song, Z.: Clustering-based performance optimization of the boiler–turbine system. IEEE Trans. Energy Convers. **23**(2), 651–658 (2008)
4. Yang, T., Liu, J., Zeng, D., et al.: Application of data mining in boiler combustion optimization. In: 2010 2nd International Conference on Computer and Automation Engineering (ICCAE), vol. 2, pp. 225–228. IEEE (2010)
5. Wenjie, Z., Chen, L.: The optimizing for boiler combustion based on fuzzy association rules. In: 2011 International Conference on Soft Computing and Pattern Recognition (SoCPaR), pp. 306–311. IEEE (2011)
6. Parsa, M., Kamyad, A.V., Sistani, M.B.N.: Combustion efficiency optimization by adjusting the amount of excess air. In: 2014 5th Conference on Thermal Power Plants (CTPP), pp. 103–108. IEEE (2014)
7. Smrekar, J., Assadi, M., Fast, M., et al.: Development of artificial neural network model for a coal-fired boiler using real plant data. Energy **34**(2), 144–152 (2009)
8. Gu, Y., Zhao, W., Wu, Z.: Online adaptive least squares support vector machine and its application in utility boiler combustion optimization systems. J. Process Control **21**(7), 1040–1048 (2011)
9. Li, G., Niu, P., Liu, C., et al.: Enhanced combination modeling method for combustion efficiency in coal-fired boilers. Appl. Soft Comput. **12**(10), 3132–3140 (2012)
10. Lv, Y., Liu, J., Yang, T., et al.: A novel least squares support vector machine ensemble model for NOx emission prediction of a coal-fired boiler. Energy **55**, 319–329 (2013)
11. Lv, Y., Yang, T., Liu, J.: An adaptive least squares support vector machine model with a novel update for NOx emission prediction. Chemometr. Intell. Lab. Syst. **145**, 103–113 (2015)
12. Tan, P., Xia, J., Zhang, C., et al.: Modeling and reduction of NOx emissions for a 700 MW coal-fired boiler with the advanced machine learning method. Energy **94**, 672–679 (2016)
13. Zhou, H., Cen, K., Fan, J.: Multi-objective optimization of the coal combustion performance with artificial neural networks and genetic algorithms. Int. J. Energy Res. **29**(6), 499–510 (2005)
14. Wu, F., Zhou, H., Ren, T., et al.: Combining support vector regression and cellular genetic algorithm for multi-objective optimization of coal-fired utility boilers. Fuel **88**(10), 1864–1870 (2009)
15. Song, J., Romero, C.E., Zheng, Y., et al.: Improved artificial bee colony-based optimization of boiler combustion considering NOX emissions, heat rate and fly ash recycling for on-line applications. Fuel **172**, 20–28 (2016)
16. Zheng, W., Wang, C., Yang, Y., et al.: Multi-objective combustion optimization based on data-driven hybrid strategy. Energy **191**, 116478 (2019)

17. Chen, N.: Industrial diagnosis, material design and intelligent data base. Science (3), 53–56 (1997). (in Chinese)
18. Qiu, G., Qin, P., Chen, N.: Study on the intelligent data base for production processes of alumina industry. Comput. Appl. Chem. (3), 211–212 (1996). (in Chinese)
19. Abel, M., Silva, L.A.L., Ros, L.F.D., et al.: PetroGrapher: managing petrographic data and knowledge using an intelligent database application. Expert Syst. Appl. **26**(1), 9–18 (2004)
20. Nihalani, N., Silakari, S., Motwani, M.: Integration of artificial intelligence and database management system: an inventive approach for intelligent databases. In: 2009 First International Conference on Computational Intelligence, Communication Systems and Networks, pp. 35–40. IEEE (2009)
21. He, W., Shao, C.: The development and challenges of industrial big data analysis technology. Inf. Control **47**(04), 18–30 (2018). (in Chinese)
22. Chu, J.: Future development of intelligent plants process industry. Sci. Technol. Rev. **36**(21), 23–29 (2018). (in Chinese)
23. Chai, T.-Y.: Development directions of automation science and technology. Acta Automatica Sinica **44**(11), 5–12 (2018). (in Chinese)
24. Bestavros, A., Lin, K.J., Son, S.H., et al.: Real-Time Database Systems: Issues and Applications. Springer, Heidelberg (2012). https://doi.org/10.1007/978-1-4615-6161-3
25. Ulusoy, Ö.: Research issues in real-time database systems: survey paper. Inf. Sci. **87**(1–3), 123–151 (1995)
26. Berkeley DB. https://www.oracle.com/database/technologies/related/berkeleydb.html

Threshold Functional Dependencies
for Time Series Data

Mingyue Ji[1], Xiukun Wei[2], and Dongjing Miao[1(✉)]

[1] Harbin Institute of Technology, Harbin 150001, Heilongjiang, China
jimy2116@mails.jlu.edu.cn, miaodongjing@hit.edu.cn
[2] Jilin University of Science and Technology, Changchun 130012, Jilin, China
weixk2116@mails.jlu.edu.cn

Abstract. This paper extends traditional Functional Dependencies (FDs) to Threshold Functional Dependencies (TFDs) for Time Series Database according to the characteristics of attribute values changing rapidly by time from sensors. In contrast to the *unique-to-same* pattern in relational schema, TFDs allow determined attribute value within a certain range rather than a clear value when corresponding to the same deciding party. We find that TFDs capable of not only detecting errors resulting from attribute value out-of-bounds in one tuple horizontally, but also from a column of single attribute among several tuples vertically. And we focus more on the former in this article. We draw a clear line between FDs and TFDs because they have some intersection. And we classify TFDs for convenience of research. We provide an inference system for classified TFDs analogous to Armstrong's axioms, prove its soundness and completeness and explain their differences and connections. We perform some experiments to show effects of TFDs which make some contributions to data quality for Time Series Database.

Keywords: Time series · Functional dependency · Threshold

1 Introduction

Time Series Database (TSDB) is mainly applied to industrial monitoring such as electric power, petroleum, chemical industry, etc. It is adept at processing constantly updated and rapidly changing data and transaction with time limits. There are inevitable errors in these sensor readings. Integrity constraints in TSDB differ from relational database mainly on functional dependencies. If we say attribute values in relational database are mostly enum types, then attribute values in TSDB are mostly continuous data in their domain but are recorded discretely. And functional dependency like A → B in the former is equality because every B value determined by the unique A must equal, while the latter focus on inequality where B can change within a reasonable range in the same condition. These differences are enough to give us a motivation to create a new setup FDs for TSDB.

Variants of FDs have arisen on a small but useful and promising scale which provide more strict or lighter constraints among attributes in the previous research. Wenfei Fan

© Springer Nature Switzerland AG 2020
Y. Nah et al. (Eds.): DASFAA 2020 Workshops, LNCS 12115, pp. 164–174, 2020.
https://doi.org/10.1007/978-3-030-59413-8_14

[1] et al. propose conditional functional dependencies (CFDs) aiming at capturing the consistency of data with satisfaction of certain attribute's value as a constant. The formula is $(X \rightarrow Y, T_p)$ where T_p is the pattern keeping certain attribute constant. Flip Korn [2] et al. define FD probabilistic, approximate constraints (FDPACs) for network traffic database by converting $X \rightarrow Y$ to $f(x) \rightarrow g(x)$. It achieves by allowing an aggregate f over a set of attribute values in X to functionally determine the similar aggregate g over those in Y. R. Haux and U. Eckert [3] introduce non-deterministic functional dependencies (NFDs) for inherent connections among attributes that can be emphasized by time. For example, one patient's weight varies with hormone level at several examination dates denoting that the patient's ID determines his non-deterministic and random weight variable. It can be expressed as $ID \rightarrow \Phi(Weight)$. So NFDs can be regarded as stochastic extensions of traditional FDs. NFDs make some difference among the above state-of-art because of correlation with time, which relates to TSDB.

Inspired by the above variants of FDs, we bring the concept of our special FDs for TSDB. When applying FD $X \rightarrow Y$ to TSDB, we find that each attribute value in Y can come from a reasonable range as t_1, t_2 and t_3 in Table 1 show with A determining B's range. We can see that t_2 is correct while t_1 and t_3 violate the range. These ranges can depend on single attribute or combined attributes restricted by nature or machinery (e.g., outdoor temperature in different places), statistic analysis methods (e.g. linear or polynomial regression), sampling frequency and so on. The range maybe includes random (discrete) variables, functions or even a constant with zero-range. Regardless of types of this range, we focus more on logic on the data schema level rather than the upper statistics methods. We name this kind of FDs Threshold Functional Dependencies (TFDs). Threshold literally seems to resemble domain integrity but they are different. We will discuss it later. As shown by the Arrow 1 hinting in Table 1, we say TFDs have function of horizontal constraints embodying in rows in Table 1.

Table 1. An instance for threshold FD

	Time	A	Arrow 1 B
t_1	10:01	5.1	$4.8 \notin [5.1-1, 5.1+1]$
t_2	10:02	5.2	$5.3 \in [5.2-1, 5.2+1]$
t_3	10:03	5.6	$11.0 \notin [5.6-1, 5.6+1]$
t_4	10:04	50.4	$49.8 \in [50.4-1, 50.4+1]$
t_5	10:05	6.6 Arrow 2	$5.9 \in [6.6-1, 6.6+1]$

From other perspectives to see errors in TSDB, outlier (or anomaly) detection [4] and repairing have been studied well in TSDB. The main ideas are based on techniques such as AR model [5], Markov models [6], neural network [7] and so on. Clearly, these statistical techniques reflect the invisible constraints by time on single attribute values in a column in a table. As shown by the Arrow 2 hinting in Table 1, we name it vertical constraint. Obviously, time stamp itself is a typical vertical constraint and Shaoxu Song

[8] et al. have studied the time stamp repairing problem. These statistical techniques above can detect t_4 which is an abrupt change among t_1 to t_5 in Table 1.

Actually, not only can our TFDs express horizontal but also vertical constraints when we use statistical techniques above to restrict different values or their ranges of a column of single attribute changing with time. In this article, we focus more on TFD's horizontal constraints rather than its vertical constraints.

Proposing a new variant of traditional functional dependencies named TFDs according with detecting errors in time series database is our first contribution. The capacity of detecting TFDs covers horizontal and vertical constraints, which is our second contribution. The third contribution is classifying TFDs and presenting a sound and complete inference system for classified TFDs analogous to Armstrong's axioms. The last contribution is that the experiments show the performance of TFDs when detecting errors in time series database.

2 Threshold Functional Dependencies

Here, we will define TFDs. Consider a Time Series Database T and schema R over a set of attributes, denoted by attr(R). And we denote attributes in R as $A, B, C\dots$ and sets of attributes of $X, Y, Z\dots$

2.1 Definition

Definition 1. A Threshold FD over R is the form $X \rightarrow \Gamma(Y)$ where Γ represents the threshold of every attribute's values in Y.

We refer to X as the left-hand side, or *lhs* for short, and to Y as the right-hand side, or *rhs* for short. If one tuple t satisfies a TFD Δ in horizontal constraints, that means attribute values in Y are within the Γ range, denoted by $t \vdash \Delta$. If every tuple in table T satisfies each element in TFD set Λ in horizontal constraints, we say $T \vdash \Lambda$.

Threshold Γ represents value ranges for *rhs* decided by *lhs* with upper and lower bounds in one or several intervals. We denote these bounds as *threshold functions*. Though we have mentioned we don't want to care more about the format of *threshold functions*, we need examples to illustrate TFDs as follows. **e.g.1**, $A, B \rightarrow \Gamma(C, D)$, $\Gamma = \{C \in (f_1(A), f_2(A)], D \in [f_3(A, B), f_4(A, B)), D \in [f_5(A, B), f_6(A, B))\}$, and f_1 to f_6 are *threshold functions*. They can be implicit or explicit functions for some attribute.

Assuming A_e is an expression only containing an arbitrary attribute A from $X \cup Y$, if A_e can be expressed by $(X \cup Y)\backslash A$ explicitly, we denote A as A-*ex*. **e.g.2**, if $A \rightarrow \Gamma(B)$, $\Gamma = \{B \in [-A - A^2, A + A^2] \, (A \geq 0 \text{ and } B \text{ is integer})\}$, we have B-*ex* and A-*ex*. If A_e is trapped in the implicit *threshold functions*, we denote A as A-*im*. **e.g.3**, if $A, B \rightarrow \Gamma(A, B)$, $\Gamma = \{g_1(A, B) \leq 10\}$, $g_1 = A^3 - \sqrt{A} + A^2B^3 - A\sqrt{B}$, we have A-*im* and B-*im*. Specially we call the attribute set AB in **e.g.3** as AB-*ex* with A-*im* and B-*im*, and call attribute set CD in **e.g.1** as CD-*ex* with C-*ex* and D-*ex*. We can see if an attribute set C is C-*ex*, then arbitrary element A of C can be A-*ex* or A-*im*. Formally speaking, an attribute set C is C-*ex* if and only if C can be expressed by attributes except any element of C or can be expressed only by C itself as **e.g.3** shows. So when a proper subset S of C is S-*ex*, then there must exist $C\backslash S$-*ex* according to symmetry.

It is not necessary that everyone of *lhs* must participate in decision on *rhs*, as *C*'s range in **e.g.1** shows, in which case we denote *B* as *C-irre*. If some attribute from *lhs* is irrelevant to everyone of *rhs*, we say the attribute is *Y-irre*. In $X \rightarrow \Gamma(Y)$, we don't allow *X* are all *Y-irre*. In other words, there at least exists one attribute which is not *Y-irre*. Sometimes there is exactly no mathematical formula between *lhs* and *rhs* due to multi-implicit, but they do have some connections. We denote this situation as *vague threshold*, and this TFD as false-TFD. We will give an example of *vague threshold* later. If *X* determine *Y* by *vague threshold*, we don't consider them into the following research in this article.

2.2 Classifying

We classify TFD $X \rightarrow \Gamma(Y)$ into two cases according to the kinds of Γ.

Case 1 Γ does not Function
$X \rightarrow \Gamma(Y)$ naturally becomes $X \rightarrow Y$, which is exactly a traditional FD, as hinted by Fig. 1(a), which is called *unique-to-same*. And the threshold is zero, denoting as *empty-threshold*. Data consistency guarantees that tuples with every attribute in common value in *X* must have the same value set of *Y* where each attribute value in *Y* has a countably infinite domain. **e.g.4**, $A \rightarrow \Gamma(A)$, $\Gamma = \{A \in [A - 0, A + 0]\}$. We classify the case of **e.g.4** into TFDs.

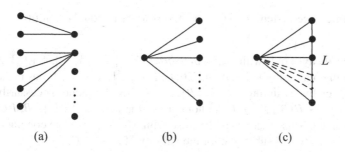

Fig. 1. Sketch showing patterns of FDs and TFDs

We hereby declare though from the perspective of concept and definition, TFDs contains part of traditional FDs, we won't mix them into one Γ in this article for the convenience of exploiting TFDs. So now our research scope is TFD including the case of **e.g.4**.

Case 2 Γ Maps More Values for Everyone in *Y* in the Same *lhs* Situation
The dependency becomes more slack relative to case 1. Tuples with every attribute in common value in *X* which have different value sets of *Y* remain existing when performing data cleaning. The value domain of every attribute in *Y* may be countably infinite or finite, as hinted by Fig. 1(b), and as *B*'s threshold in **e.g.2** shows. It may be non-countably continuous, as hinted by Fig. 1(c) where the nodes can slide along the straight line *L* if we allow *B* not must be integrity in **e.g.2**. Specially, when $A = 0$ in **e.g.2**, though *B* can only be zero, it is different from case 1 because of dynamic variability of *A*.

2.3 Properties of TFDs

Here, we describe the special characteristics of TFDs. At the same time we explain the differences and connections between TFDs and traditional FDs, also between TFDs and domain constraints.

- The consistency restricted by traditional FDs must be reflected in more than two tuples. But one single tuple can accord with or violate a TFD in horizontal constraints.
- Every traditional FD set can have its corresponding consistent table before or after repairing. So data schema over traditional FDs is always worth repairing but it is not the case over TFDs. Not every TFDs set Λ can make sense due to the range of *rhs*. **e.g.5**, $A \rightarrow \Gamma_1(C), \Gamma_1 = \{C \in [1/A, +\infty] (A > 0)\}$ and $B \rightarrow \Gamma_2(C), \Gamma_2 = \{C \in [-\infty, -1/B] (B > 0)\}$. We can see there doesn't exist one correct C value at all.
- As mentioned earlier, TFDs appear literally to be like domain constraints but they are totally different. The former imposes restrictions to certain column data determined by other attributes while the latter limits every column data and has nothing to do with other attributes.

3 An Inference System for TFDs

3.1 TFD Rules

There is an inference system for TFD analogous to Armstrong's Axioms.

Lemma 1. $A \rightarrow \Gamma(A)$

   This TFD always holds with *empty-threshold* $\Gamma = \{A \in [A - 0, A + 0]\}$. $A \rightarrow \Gamma(A)$ makes sense in both traditional FDs and TFDs. In case 1, $A \rightarrow \Gamma(A)$ has been classified into TFD. Specially, assuming $X = \{A, B\}$, $X \rightarrow \Gamma(X)$ holds not necessarily due to $A \rightarrow \Gamma(A)$ and $B \rightarrow \Gamma(B)$. $X \rightarrow \Gamma(X)$ holds may due to $\Gamma = \{A \in [-B, B], B \in [-A, A]\}$. Moreover, **e.g.3** also belongs to this case with *AB-ex*. So we can conclude that $A \rightarrow \Gamma(A)$ is not necessity and substitute for the validity of $X \rightarrow \Gamma(X)$.

TFD1. If $Y \subseteq X \subseteq R$, then $X \rightarrow \Gamma(Y)$.
TFD1 corresponds to Armstrong's reflexivity.
TFD2. If $X \rightarrow \Gamma_1(Y)$ and $Z \subseteq R$, then $XZ \rightarrow \Gamma_1(Y)$ and $XZ \rightarrow \Gamma_2(YZ)$.
TFD2 extends Armstrong's augmentation. This rule emphasizes not every one of *lhs* must determine *rhs*. Actually these redundant items are encouraged to be cleaned.
TFD3. If $X \rightarrow \Gamma_1(Y)$ and $Z \rightarrow \Gamma_2(W)$, then $XZ \rightarrow \Gamma_3(YW)$.
TFD3 also extends Armstrong's augmentation.
TFD4. If $X \rightarrow \Gamma_1(Y)$, and $Z \subseteq Y$ with *Z-ex*, then $X \rightarrow \Gamma_2(Z)$.
TFD4 is similar to the decomposition corollary in Armstrong's axioms.
TFD5. If $XY \rightarrow \Gamma(Z)$, X and Y affect Z respectively, then $X \rightarrow \Gamma_1(Z)$ and $Y \rightarrow \Gamma_2(Z)$.
TFD6. If $XY \rightarrow \Gamma(Z)$ with $W \subseteq XY$, and W doesn't work on Z range, then $XY \backslash W \rightarrow \Gamma(Z)$.

For TFD5 and TFD6, there are no counterpart axioms for Armstrong's traditional FDs. And they are complementary to TFD2. If we apply TFD5 and TFD6 to a TFD, **e.g.1**, $A, B \to \Gamma(C, D)$, $\Gamma = \{C \in (f_1(A), f_2(A)], D \in [f_3(A, B), f_4(A, B)), D \in [f_5(A, B), f_6(A, B))\}$, then we can obtain by TFD5 $A, B \to \Gamma_1(C)$, $\Gamma_1 = \{C \in (f_1(A), f_2(A)]\}$ and $A, B \to \Gamma_2(D)$, $\Gamma_2 = \{D \in [f_3(A, B), f_4(A, B)), D \in [f_5(A, B), f_6(A, B))\}$. And by TFD6, the former one becomes $A \to \Gamma_3(C)$, $\Gamma_3 = \{C \in (f_1(A), f_2(A)]\}$.

Specially, TFD2, TFD5 and TFD6 are contrary to Armstrong's merger and decomposition corollaries, servicing for *lhs* and *rhs* respectively.

TFD7. If $X \to \Gamma(Y)$ with $Ram \subseteq X \cup Y$, and Ram is Ram-ex, then $X \cup Y \to \Gamma_i(Ram)$.

There is not an issue for traditional FDs corresponding to TFD7. Actually TFDs mean inequalities, so some attributes can become independent variables in *threshold functions* with others becoming dependent variables. The dependent variables are exactly the *rhs*. It seems like pulling one hair and moving the whole body.

In TFDs, we don't have Armstrong's transitivity counterpart. Without loss of generality, transitivity in TFDs will bring *vague threshold*. The illustration is as follows. If one tuple t satisfies $X \to \Gamma_1(Y)$ and $Y \to \Gamma_2(Z)$, then we know that $t[Z]$ is within the Γ_2 range decided by $t[Y]$ which is decided by $t[X]$. So in the same tuple, $t[X]$ affects $t[Z]$'s value.

But there doesn't always exist a non-false-TFD between X and Z due to implicit bridge Y. **e.g.6**, $A \to \Gamma_1(B)$, $\Gamma_1 = \{B \in [A - 10, A \ 10]\}$, $B \to \Gamma_2(C)$, $\Gamma_2 = \{C \in [B, + \infty]\}$, so we can deduce $A \to \Gamma_3(C)$, $\Gamma_3 = \{C \in [A + 10, + \infty]\}$. And **e.g.7**, if we change the first TFD above into $A \to \Gamma_1(B)$, $\Gamma_1 = \{B * \ln B + 1/B \in [A - 10, A + 10]\}$, then the last TFD above comes to $A \to \Gamma_3(C)$ with *vague threshold* Γ_3, and this is the case of false-TFD. Sometimes we attain a TFD by transitivity as **e.g.6** shows, but the result is false-TFD more often as **e.g.7** shows. For convenience and unity, we don't accept transitivity of TFDs.

3.2 Sound and Complete

Here we prove soundness and completeness of the inference system.

Definition 2. Let Λ be a set of TFDs. If $X \to \Gamma(Y)$ is deduced from Λ using TFD rules above, then we say $\Lambda \vDash X \to \Gamma(Y)$.

Definition 3. Let Φ be a set of inference rules {TFD1 to TFD7}. Then Φ is sound for logical implication of TFDs if $X \to \Gamma(Y)$ is deduced from Λ using Φ, and $X \to \Gamma(Y)$ is true in any relation in which the TFDs of Λ are true.

Lemma 2. Φ is sound for logical implication of TFDs.

Proof. In order to prove the soundness of Φ we have to prove that each of the TFD rules is sound. The processes are Proof 1 to Proof 7.

Proof 1. If $Y \subseteq X \subseteq R$, then $X \to \Gamma(Y)$.

For each attribute in Y, denoted as Ram, $Ram \to \Gamma(Ram)$ holds according to Lemma 1. $Ram \to \Gamma(Ram)$ guarantees that when $X \backslash Ram$ don't determine Ram, *there* at least exists one *threshold function* for *Ram* to hold for Γ. **e.g.8**, $AB \to \Gamma(AB)$, $\Gamma = \{A \in [A - 0, A + 0]$, $B \in [B - 0, B + 0]\}$. When $X\backslash Ram$ decide *Ram's* values by some *threshold functions*, $Ram \to \Gamma(Ram)$ is right but doesn't work in Γ as we discuss in Lemma 1. **e.g.9**, $AB \to \Gamma(AB)$, $\Gamma = \{A \in [B - 1, B + 1]$, $B \in [A - 1, A + 1]\}$.

Proof 2. If $X \to \Gamma_1(Y)$ and $Z \subseteq R$, then $XZ \to \Gamma_1(Y)$ and $XZ \to \Gamma_2(YZ)$.

If one tuple t satisfies $X \to \Gamma(Y)$, then $t[Y]$ is within the Γ range decided by $t[X]$, also by $t[XZ]$ considering Z as Y-irre. So $XZ \to \Gamma(Y)$ holds with *threshold functions* unchanged. In the same tuple, $t[Z]$'s value range also can be determined by $t[Z]$ exactly as Lemma 1 shows, considering X as Z-irre. So $XZ \to \Gamma(YZ)$ holds with added *threshold functions* like in **e.g.8**.

Proof 3. If $X \to \Gamma_1(Y)$ and $Z \to \Gamma_2(W)$, then $XZ \to \Gamma_3(YW)$.

If one tuple t satisfies $X \to \Gamma_1(Y)$ and $Z \to \Gamma_2(W)$, then $t[Y]$ is within the Γ_1 range decided by $t[X]$ and $t[W]$ is within the Γ_2 range decided by $t[Z]$. In the same tuple, we can say $t[Y]$ and $t[W]$ are within respective ranges decided by $t[X]$ and $t[Z]$ combined in Γ_3. So $XZ \to \Gamma_3(YW)$ holds.

Proof 4. If $X \to \Gamma_1(Y)$, and $Z \subseteq Y$ with Z-ex, then $X \to \Gamma_2(Z)$.

If one tuple t satisfies $X \to \Gamma_1(Y)$ and $Z \subseteq Y$ with Z-ex, we can convert it into $X \to \Gamma_1(Z \cup (Y\backslash Z))$. In tuple t, we can say $t[Z]$ and $t[Y\backslash Z]$ are both within the Γ_1 range decided by $t[X]$. We divide *threshold functions* in Γ_1 into Γ_2 and Γ_3 for Z and $Y\backslash Z$ respectively. So we can say in the same tuple t, $t[Z]$ is within the Γ_2 range decided by $t[X]$ while $t[Y\backslash Z]$ is within Γ_3. So $X \to \Gamma_2(Z)$ and $X \to \Gamma_3(Y\backslash Z)$ hold.

Proof 5. If $XY \to \Gamma(Z)$, X and Y affect Z respectively, then $X \to \Gamma_1(Z)$ and $Y \to \Gamma_2(Z)$.

If one tuple t satisfies $XY \to \Gamma(Z)$, X and Y affect Z range respectively, then $t[Z]$ is within the Γ range decided by $t[X]$ and by $t[Y]$ respectively. So we can say $X \to \Gamma_1(Z)$ and $Y \to \Gamma_2(Z)$ hold. **e.g.10**, if $A, B \to \Gamma(C)$, $\Gamma = \{C \in [f_1(A), f_2(A)], C \in [f_3(B), f_4(B)]\}$, then $A \to \Gamma_1(C)$, $\Gamma_1 = \{C \in [f_1(A), f_2(A)]\}$ and $B \to \Gamma_2(C)$, $\Gamma_2 = \{C \in [f_3(B), f_4(B)]\}$ hold.

Proof 6 If $XY \to \Gamma(Z)$ with $W \subseteq XY$, and W doesn't work on Z range, then $XY\backslash W \to \Gamma(Z)$.

If one tuple t satisfies $XY \to \Gamma(Z)$ with $W \subseteq XY$, and W doesn't work on Z range, then $t[Z]$ is within the Γ range decided by $t[XY\backslash W]$. So $XY\backslash W \to \Gamma(Z)$ holds. **e.g.11**, if $A, B \to \Gamma(C)$, $\Gamma = \{C \in [f_1(A), f_2(A)]\}$, then we have $A \to \Gamma(C)$, $\Gamma = \{C \in [f_1(A), f_2(A)]\}$.

Proof 7. If $X \to \Gamma(Y)$ with $Ram \subseteq X \cup Y$, and Ram is Ram-ex, then $X \cup Y \to \Gamma_i(Ram)$.

If one tuple t satisfies $X \to \Gamma(Y)$, then $t[Y]$ is within the Γ range decided by $t[X]$. Denoting some elements in $X \cup Y$ as Ram, if Ram is Ram-ex, then $t[Ram]$'s value is affected by attributes $X \cup Y\backslash Ram$ or Ram itself like in Lemma 1. So we have $X \cup Y \to$

$\Gamma_i(Ram)$ due to permissible redundant of *lhs* implemented by TFD2. **e.g.12**, $A, B \rightarrow \Gamma_1(C)$, $\Gamma_1 = \{C \in [A + B - 10, A + B + 10]\}$, so $A, C \rightarrow \Gamma_2(B)$, $\Gamma_2 = \{B \in [C - A - 10, C - A + 10]\}$ and $B, C \rightarrow \Gamma_3(A)$, $\Gamma_3 = \{A \in [C - B - 10, C - B + 10]\}$. Or we have **e.g.3** as example. But when *Ram* is not *Ram*-ex, **e.g.13**, $A, B \rightarrow \Gamma_1(C)$, $\Gamma_1 = \{C \in [B + e^{AB} + A - 10, B + e^{AB} + A + 10]\}$, then we cannot get $A, C \rightarrow \Gamma_2(B)$ or $B, C \rightarrow \Gamma_3(A)$ but $A, B \rightarrow \Gamma(C)$ is correct.

Definition 4. Let Φ be a set of inference rules {TFD1 to TFD7}. Let $\Lambda +$ be all TFDs which can be deduced by Λ using Φ and Λ itself. Φ is complete if every element of Λ + can be deduced by starting from Λ and reasoning from Φ.

It is NP-hard to find out all TFDs in $\Lambda+$ as in Armstrong's axioms for traditional FDs. So we prove TFD rules' completeness by contrapositive way.

Proposition 1. If $X \rightarrow \Gamma(Y)$ can never be deduced from Λ by using Φ, then $X \rightarrow \Gamma(Y)$ can never be implicated in $\Lambda+$.

Lemma 3. The inference rules TFD1 to TFD7 are complete.

Proof. Contrapositive of TFDs rules' completeness, that is Proposition 1, proves Lemma 3.

Theorem 1. The inference system is sound and complete.

4 Experiments

In this section, we present results of detecting TFDs horizontal violations over climate data by algorithm shown in Fig. 2.

Algorithm *Detecting_horizon*
Input: a set of TFDs Λ,a TSDB table T
Output: tuples err[n] with error data
1.for each TFD Δ in Λ **do**
2.   Remove irrelevant attributes to *rhs* in *lhs* by TFD6.
3.end for
4.for each tuple t in T **do**
5.   **for** each TFD Δ in Λ **do**
6.   **for** each formula Γ in Δ **do**
7.   attr[R]←attributes in Γ.
8.   **if** attr[R] does not satisfy Γ **then**
9.   err[n]←t.
10.   **end if**
11.   **end for**
12.   **end for**
13.end for

Fig. 2. Algorithm detecting horizontal errors

We find a TFD between Sea Level Pressure (SLP) and elevation in climate series data based on (1) where p represents SLP/Pa and h represents elevation/m.

$$p = e^{5.25885 \times \ln(288.15 - 0.0065h) - 18.2573} \tag{1}$$

So from (1) we obtain a TFD $h \to \Gamma(p)$, $\Gamma = \{p \in [(1) - \delta, (1) + \delta]\}$. Though the above equality is deduced in the condition of considering influence of atmosphere and temperature, there must exist a confidence interval in natural circumstances. In our experiments, we let δ be 35 and intuitive detecting-errors results are as Fig. 3(a), (b), (c) show.

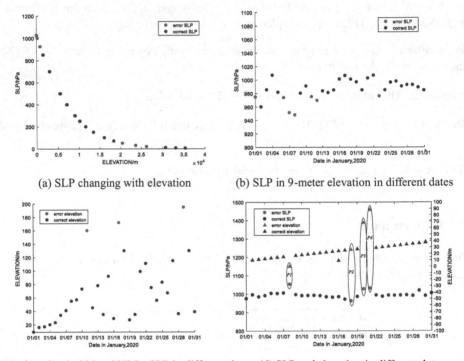

(a) SLP changing with elevation   (b) SLP in 9-meter elevation in different dates

(c) elevation in 996 to 997hPa SLP in different dates (d) SLP and elevation in different dates

Fig. 3. Experimental results

It is worth mentioning that we execute the algorithm in [9] which can detect abrupt changes with time to obtain errors among tuples while performing *detecting_horizon* algorithm, for which Fig. 3(d) shows horizontal and vertical constraints at the same time. Time node P_1 doesn't satisfy both constraints while P_2 satisfies the latter but not the former and P_3, P_4 are exactly opposite to P_2.

From the above results, we can see that there are several types of errors occurred in TSDB, those satisfy or don't satisfy TFDs with or without abrupt changes. If one attribute A is an abrupt change in a column data, then other attributes in TFD including A are greatly possibly abrupt changes because of value bindings in TFD.In this case, we just need to check vertical constraints only in the column of attribute A and its corresponding TFDs to save time and energy. So TFDs reduce the burden of vertical detection of every column data, as Fig. 4 shows, and TFDs can detect errors between attributes in one tuple that cannot achieved in vertical detection.

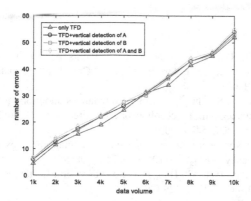

Fig. 4. Detecting errors in different settings of techniques

If we classify types of errors in TSDB with the same probability of each error type by satisfying or not satisfying with or without abrupt changes, we can obtain $2^{n+1} - 1$ errors when there are n attributes in one TFD. Under the condition of implementation of the single TFD with n attributes, the rates of error types that can be detected by arbitrary number of attributes' vertical detection are shown as Table 2.

Table 2. The rates of error types that can be detected

Error types	Only detecting TFD	Detecting TFD and vertical constraints in one attribute	Detecting TFD and vertical constraints in two attributes	Detecting TFD and vertical constraints in k attributes ($k \leq n$)
$2^{n+1} - 1$	$2^n/2^{n+1} - 1 \geq 50\%$	$2^n + 2^{n-1}/2^{n+1} - 1 \geq 75\%$	$2^n + 2^{n-1} + 2^{n-2}/2^{n+1} - 1 \geq 87.5\%$	$2^n + 2^{n-1} + ...+ 2^{n-k}/2^{n+1} - 1$

Actually, numbers of errors satisfying TFD with abrupt changes such as P_3 an P_4 in Fig. 3(d) are less than other types, so rates in the above table will be arose in practice.

5 Conclusions

We propose a new variant of traditional Functional Dependencies in this article and provide classified TFDs with some rules for its logical operation. But the theoretical system is not that perfect. We have mentioned the most different content contrasting to traditional FD's system is the correctness of TFDs set. So next we will focus on this problem and find effective algorithm to judge correctness of a TFD set. In addition, the vertical constraint hidden in TFDs has been still unsolved though we simply show its capacity to detect errors in the experiments. It will be a key point of our future work.

References

1. Bohannon, P., Fan, W., Geerts, F., Jia, X., Kementsietsidis, A.: Conditional functional dependencies for data cleaning. In: IEEE 23rd International Conference on Data Engineering, ICDE 2007, pp. 746–755 (2007)
2. Korn, F., Muthukrishnan, S., Zhu, Y.: Checks and balances: monitoring data quality problems in network traffic databases. In: Proceedings of 29th International Conference on Very Large Data Bases, VLDB, pp. 536–547 (2003)
3. Haux, R., Eckert, U.: Non deterministic dependencies in relations: an extension of the concept of functional dependency. Inf. Syst. 2(10), 139–148 (1985)
4. Gupta, M., Gap, J., Aggarwal, C., Han, J.: Outlier detection for temporal data. Synth. Lect. Data Min. Knowl. Discov. 5(1), 1–129 (2014)
5. Zhang, A., Song, S., Wang, J., Philip, Y.: Time series data cleaning: from anomaly detection to anomaly repairing. PVLDB 10(10), 1046–1057 (2017)
6. El Chamie, M., Janak, D., Açıkmeşe, B.: Markov decision processes with sequential sensor measurements. Automatica 103, 450–460 (2019)
7. Kolanowski, K., Swietlicka, A., Kapela, R., Pochmara, J., Rybarczyk, A.: Multisensor data fusion using Elman neural networks. Appl. Math. Comput. 319, 236–244 (2018)
8. Song, S., Gao, Y., Wang, J.: Cleaning timestamps with temporal constraints. PVLDB 9(10), 708–719 (2016)
9. Jordan, H., Owen, S.V., Arun, K.: Automatic anomaly detection in the cloud via statistical learning. CoRR abs/1704.07706 (2017)

A Survey on Modularization of Chatbot Conversational Systems

Xinzhi Zhang, Shiyulong He, Zhonfei Huang, and Ao Zhang$^{(\boxtimes)}$

College of Intelligence and Computing, Tianjin University, Tian Jin, China
azhang@tju.edu.cn

Abstract. Chatbots have attracted more and more attention and become one of the hottest technology topics. Deep learning has shown excellent performance in various fields such as image, speech, natural language processing and dialogue, it has greatly promoted the progress of chatbots, it can use large amounts of data to learn response generation and feature representations. Due to the rapid development of deep learning, hand-written rules and templates were quickly replaced by end-to-end neural networks. Neural networks is a powerful model that can solve generation problems in conversation response. People's requirements for chatbots have also increased with the continuous improvement of neural network models. In this article, we discuss three main technologies in chatbots to meet people's requirements, syntax analysis, text matching and sentiment analysis, and outline the latest progress and main models of three technologies in the field of chatbots in recent years.

Keywords: Chatbots · Neural networks · Deep learning

1 Introduction

Nowadays, with the rapid development of intelligent question and answer system, chatbot has been gradually integrated into all aspects of human society. Whether in academia, industry or People's Daily lives, we can always see a variety of chatbots make outstanding contributions. In recent years, the continuous innovation and expansion of artificial intelligence, neural network and deep learning have greatly improved the functions and features of open domain chatbots. As a result, humans have the ability to communicate more smoothly with machines, and chatbots behave more like humans.

As a system for communicating directly with humans, we are also increasingly demanding of chatbots. For task-oriented robots that help us achieve a certain goal (most of which exist now), we require the accuracy and validity of the information; In contrast, most of the non-task robots used for small talk require semantic correctness while also paying attention to the rationality and consistency of response when communicating with people. In recent years, great progress has been made in the research and technology of search-based and generative chatbots, which also provides us with broader ideas for relevant exploration.

© Springer Nature Switzerland AG 2020
Y. Nah et al. (Eds.): DASFAA 2020 Workshops, LNCS 12115, pp. 175–189, 2020.
https://doi.org/10.1007/978-3-030-59413-8_15

For these models, the existing research mainly focuses on the overall technology, but the overall technology can be further divided into multiple modules, among which the most important module technologies can be divided into three: syntactic analysis, text matching and emotion analysis. Syntactic analysis is the basic work of natural language processing. It analyzes the syntactic structure of a sentence (subject-verb-object structure) and the dependencies between words (juxtaposition, subordination, etc.). Syntactic analysis can lay a solid foundation for the application scenarios of natural language processing, such as semantic analysis, affective tendency and thought extraction. Through emotion analysis, the advantages and disadvantages of products in various dimensions can be explored to determine how to improve products. Emotion analysis based on deep learning has the advantages of high accuracy, strong universality and no need of emotion dictionary. Text matching is mainly divided into "deep learning model based on single-turn response matching" and "deep learning model based on multi- turn response matching". Based on the deep learning model of single-turn response matching, the two documents to be matched are mapped to two vectors, and then the two vectors are transferred through the neural network to output the results, and the conclusion of whether they match is drawn. The feature of the deep learning model based on multi- turn response matching is that the two documents to be matched are expressed as words, phrases, sentences and other different granularity through the neural network, and then the similarity matrix is crossed and input into the neural network to obtain the conclusion whether they match or not.

The important modules are always essential, they will not always have the overall architecture of the big change, in this case, through a theoretical analysis of the techniques used in the three modules, the understanding of performance, accuracy, effect and analysis, refining the overall performance of the master in pairs, deep understanding of the structure of the various modules and the reasons of the different performance, which can be more effective from partial optimization to enhance and improve the overall performance.

In this paper, we will review and summarize relevant researches of the above three modules in recent years. In the following sections, we describe the characteristics and applications of these three modules in more detail.

2 Syntactic Analysis

Syntactic analysis is one of the core problems in language comprehension and has received extensive attention. Dependency parsing is a popular method to solve this problem, because there are dependency tree libraries in many languages (Buchholz et al. 2006; Nivre et al. 2007; McDonald et al. 2013) [1] and dependency parsers.

Chen and Manning et al. (2014) [2] proposed a neural network version of the parser based on greedy transformation. In their model, feedforward neural networks with a hidden layer are used to make transition decisions. The hidden layer has the ability to learn any combination of atomic inputs, eliminating the need for manual design features. In addition, because the neural network adopts distributed representation, the similarity between lexical markers and arc markers can be modeled in continuous space. Despite their model than the greedy manual design of similar products better, but it cannot compete with the most advanced dependency parser, which is trained for a structured search. Greed model although extremely fast, but usually run into search errors, because they are unable to recover from the wrong decision.

David Weiss and Chris Alberti et al. (2015) [3] proposed a structured perceptron training method based on dependency analysis based on neural network transformation, combined the representation ability of neural network with structured training and advanced search of reasoning support, and used a large number of sentences expanded by automatic analysis corpus to learn the neural network representation. This work started from the basic structure of Chen and Manning et al. (2014), but there was a further improvement in the architecture and optimization process: allowing smaller POS tags to be embedded and Relu units to be used in the hidden layer. These improvements improved the accuracy of the model by nearly 1% compared with Chen and Manning et al. (2014) (Fig. 1).

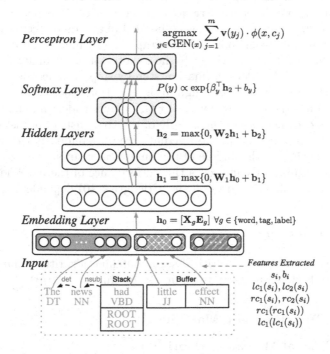

Fig. 1. Representation model framework.

In addition, they also proposed an effective method to use unmarked data, called "three-training" method. They used two different parsers to parse the unmarked data, and only selected two parsers to generate statements of the same tree, thus generating a large number of trusted parse trees.

In the same year, Mingbo Ma, Liang Huang et al. (2015) [4] proposed a very simple convolutional neural network (DCNNs) based on dependence, which is similar to the sequential CNNs model of Kim et al. (2014) [5], but the difference is that Kim et al. (2014) put a word in the sequence of the text, the model will be a word and its father, grandfather and great grandfather, and brothers and sisters in dependence on the tree, in this way to integrate information over a long distance. This is a very simple correlational convolution framework that performs better than sequential CNNs at sentence modeling.

Tao, Ji, Yuanbin Wu et al. (2019) [6] proposed a map neural network (GNNs) to learn to represent dependency tree nodes, said will map neural network is added into dependency parsing, efficient higher order information coding to rely on the representation of a tree node in: given a sentence, a parser for all the words to score in the first place, see if they can keep effective dependencies, and then use the decoder (for example, greed, maximum spanning tree) generated from these marks a complete parsing tree. Two previous outstanding works on node performance were recursive neural networks (RNNs) (Kiperwasser and Goldberg et al. (2016)) and biaffine mappings (Dozat and Manning et al. (2017)), but these representations ignore the characteristics associated with dependency structures.

Given a weighted graph, a GNN is embedded in a node by recursively aggregating the nodes of its neighbors, and the graph can be modified during parsing. The representation of a node through the superposition of multi-layer GNNs, collect all kinds of high order information step by step, into decoder with global evidence final decision with recent high approximation order parser, GNNs output calculation based on the previous layer GNN layer node, said and GNN node vector updates can check all in the middle of the tree, instead of just extracting the higher-order features of an intermediate tree, therefore, it can reduce the influence of subprime in the middle of the analytical results.

This parser significantly improves the performance of the baseline parser on long sentences, but slightly worse on short (length < 10) dependent lengths.

3   Text Matching

Text matching in chatbots is a critical step, matching algorithms must enhance the correlation between posts and responses (B. Hu, et al. 2014) [9].

3.1   Single-Turn Response Matching

3.1.1   Traditional Matching Algorithm

Early matching techniques mostly matched at the lexical level, that is, how much the query field covers to calculate the matching score between the two. The higher the score, the better the matching degree of the query. The traditional matching models mainly include BoW, VSM, BM25, SimHash, etc. Among them, the most classic models are WMD, BM25. Matt et al. (2015) [7] associates word embeddings with EMD to measure document distance. WMD (word mover 's distance) algorithm is proposed, it models the document distance as a combination of the semantic distances of the words in the two documents, such as the Euclidean distance of the word vectors corresponding to any two words in the two documents and then weighted sum. The BM25 algorithm is commonly used to search for correlation bisectors. It performs morpheme analysis on Query to generate morpheme qi; then, for each search result D, calculate the correlation score of each morpheme qi and D, and finally, weight the sum of the correlation scores of qi and D to obtain Correlation score between Query and D.

3.1.2 Deep Matching Algorithm

Matching algorithms are based on vocabulary matching, so they have great limitations. Deep learning methods can solve the problem of semantic limitations in traditional methods by extracting features and training data from the original data, and with the technology of word vector to achieve semantic level matching. In addition, based on the hierarchical structure of the neural network, the deep matching model can better establish a hierarchical matching model. Generally, deep text matching models are divided into two categories, representation model and interaction model.

Representation model focuses on the construction of the presentation layer, which transforms text into a unique overall representation vector at the presentation layer which is displayed in Fig. 2.

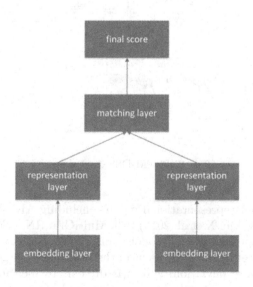

Fig. 2. Representation model framework.

The representational model is based on the Siamese network (Yann Lecun et al. 2005). The two texts are first mapped to a unified space, and the overall semantics of the text are extracted before matching. At the matching layer, the dot product, cosine, Gaussian distance, MLP, similarity matrix and other methods are used for interactive calculation. Deep Structured Semantic Models (DSSM) (Huang PS et al. 2013) is a basic representational model. It first encodes two pieces of text into a fixed-length vector, and then calculates the similarity between the two vectors. The relationship between texts, where Q is a query and D is each candidate document.

The disadvantage of DSSM is that it uses the bag-of-words model (BoW), which loses word order information and context information, and it uses a weakly supervised, end-to-end model, with unpredictable prediction results. In response to the shortcomings of the DSSM bag-of-words model losing context information, CNN-DSSM (Shen, Yelong et al. 2014) [11] emerged at the historic moment. Its difference from DSSM mainly lies in the input layer and the presentation layer. CNN-DSSM uses CNN to extract local information, and then uses max pooling to extract and summarize global information in the upper layer. It can keep the context information more effectively, but cannot keep the context information that is far apart. To address this shortcoming, LSTM-DSSM (Palangi, Hamid et al. 2014) [12] was proposed. Here is its overall network structure (Fig. 3):

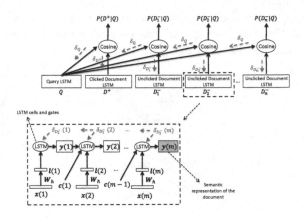

Fig. 3. LSTM-DSSM model.

In addition, other representational models including ARC-I (Hu, Baotian et al. 2015) [9], CNTN (Qiu X et al. 2015) [13], MultiGranRNN (Yin W et al. 2015) [14], and so on. Representational models can pre-process text. Constructing an index, but it will lose the semantic focus of the resulting sentence representation, are prone to semantic deviation, and it is difficult to measure the contextual importance of the word. Therefore, interaction model is proposed.

The interactive model discards the idea of post-matching. It assumes that the global matching degree depends on the local matching degree. The first matching between words is performed at the input layer, and the matching result is used as a grayscale image for subsequent modeling (Fig. 4):

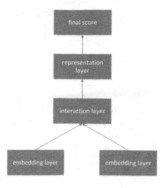

Fig. 4. Interaction model framework.

Interaction model uses the matching signal between words as a grayscale image, and then performs subsequent modeling abstraction. In the interaction layer, an interaction matrix is formed by two text words and words. In the representation layer, the interaction matrix is abstractly represented, using CNN or S-RNN (Yu, Zeping et al. 2018) [15]. Bao et al. (2014) [9] proposed the ARC-II network structure for the text matching model. Assuming that the length of both sentences is N and the embedding dimension is D, Then use a 3 * 3 convolution kernel to scan on an N * N picture, each scan 3 horizontal grids, 3 vertical grids, which respectively represent the words corresponding to two sentences, and then take 6 words, A total of 6 * D, then the size of the convolution kernel is also 6 * D. The convolution kernel and the selected word are multiplied and added (there is no activation function here), and finally a value is obtained, and the 3 * 3 volume is moved Kernel, and finally get a convolution picture (Fig. 5):

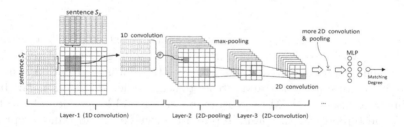

Fig. 5. ARC-II model.

In addition, Liang Pang et al. (2016) [16] proposed the MatchPyramid model, which uses image recognition to perform text matching, converts text matching to Text Matrix, and builds a CNN pyramid model to complete matching prediction. Construct matching matrices from three perspectives, consider the two-to-two relationship between words in sentences more carefully, construct three matrices for superposition, treat these matrices as pictures, and use convolutional neural networks to extract features from the matrices (Fig. 6).

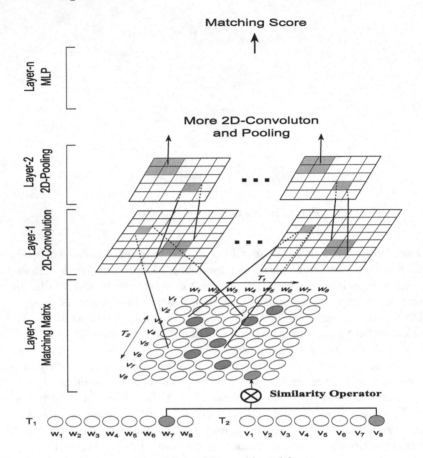

Fig. 6. MatchPyramid model.

Qian Chen et al. (2017) [17] proposed another text similarity calculation model ESIM, which consists of four parts: Input Encoding, Local Inference Modeling, Inference Composition and Prediction. Intra-sentence attention is mainly used to realize local inference, and further to achieve global inference. The Input Encoding layer uses BiLSTM for feature extraction and keeps the hidden state, where a and b represent the premise p and hypothesis h, and i and j represent different moments:

$$\bar{a}_i = BiLSTM(a, i), i \in [1, ..., l_a]$$

$$\bar{b}_j = BiLSTM(b, j), j \in [1, ..., l_b]$$

Next, calculate the similarity between the two sentences word to obtain a 2-dimensional similarity matrix, and then perform a local inference of the two sentences. Use the previously obtained similarity matrix to combine the a and b sentences. Generate similarity-weighted sentences to each other with the same dimensions.

$$\tilde{a}_i = \sum_{j=1}^{l_b} \frac{exp(e_{ij})}{\sum_{k=1}^{l_b} exp(e_{ik})} \bar{b}_j, i \in [1, ..., l_a]$$

$$\tilde{b}_j = \sum_{i=1}^{l_a} \frac{exp(e_{ij})}{\sum_{k=1}^{l_a} exp(e_{kj})} \bar{a}_i, j \in [1, ..., l_b]$$

In the Inference Composition layer, the previous value is sent to the BiLSTM again to capture the local inference information and context for inference combination. Finally, the tanh activation function is used to send it to the Softmax layer.

Sentences should not only consider the direction from the question to the answer, but should also infer the question from the answer. Zhiguo Wang et al. (2017) [18] proposed the BiMPM model. Its innovation lies not only in the bidirectionality, but also in the consideration of sentences. There are 4 different ways to interact with each other, and then all the results are stitched and predicted. Word Representation Layer, representing each word in the sentence as a d-dimensional vector. Context Representation Layer fuses the context information into the representation of each time-step of P and Q. The output of the Matching Layer is two sequences. Each vector in the sequence is a certain time-step of a sentence matches all the time-steps of another sentence. Aggregation Layer aggregates two sequences of matching vectors into a fixed-length matching vector (Fig. 7).

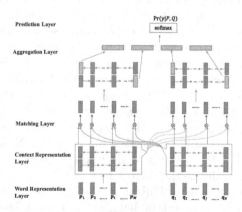

Fig. 7. BiMPM model.

In addition, typical interaction models also include DeepMatch, ABCNN (Wenpeng Yin et al. 2016) [19], DeepRank (Liang Pang et al. 2017) [20], IR-Transformer, and so on.

3.2 Multi-turn Response Matching

Interaction model can better grasp the semantic focus and better model the contextual importance, but it ignores global information such as syntax and cross-sentence comparison, and cannot describe global matching information from local matching information. And deep text matching is a single-round match question and answer for retrieval dialogues, which cannot complete multiple rounds of tasks. Therefore, based on a single round, adding other feature extraction techniques can complete multiple rounds of dialogue.

Xiangyang Zhou et al. (2016) [21] proposed to merge multiple rounds of question-and-answer sentences into one column, separated by _SOS_ at the connection, regarding the entire conversation history as "one sentence" to match the next sentence, and merge the entire conversation history into one column. After doing word embedding, lexical-level features are extracted through the GRU module and matched with candidate responses. In addition, the article proposes to match each text once, which means the combination of word-level and utterance-level. The loss function used in the integrated model is disagreement-loss (LD) and like-hood-loss (LL), it is an extension of a single-round question-and-answer representation model.

SMN was proposed by Yu Wu et al. (2017) [22], it is an extension of the single-turn Q & A interaction model. Constructing interactive representations of historical questions and answers and candidate replies is important feature information, so we use the matching matrix in semantic matching to construct a model by combining CNN and GRU. Here we consider the similarity matrix of two texts as an image, and then use the image classification model CNN to obtain higher level similarity feature representations (such as phrase level, segment level, etc.), and finally obtain the global similarity matching feature. Both the Multi-view and SMN models treat the conversation history as a whole. This will ignore the internal characteristics of the conversation history. For example, a conversation often includes multiple topics. In addition, the importance of words and sentences in a conversation is also different. Aiming at the information characteristics in these dialogue histories, a DUA model (Zhousheng Zhang et al. 2018) [23] was proposed. The author believes that the last sentence in the historical dialogue information is usually the most critical. Then make self-attention with utterance as the unit, the purpose is to filter redundancy (such as meaningless empty words) and extract key information. DUA focuses on multi-level feature extraction for both vocabulary and sentences, but in the case of many rounds, it will misjudge the correct response.

In order to solve the problem of misjudgment of response, the attention mechanism is applied to multiple rounds of dialogue. Xiangyang Zhou et al. (2018) [24] used self-attention and cross-attention to extract response and context features, which are mainly divided into Representation, Matching and Aggregation. In the matching stage, the DAM model has two matching matrices, which are the core part of the entire model. The first column Mself is called self-attention-match, and the second column Mcross is called cross-attention-match. Mself consists of Ui and R composed of L matrices obtained by Ui and response in each sentence of the previous layer. Mcross uses two layers of traditional attention to calculate the alignment matrix.

Chongyang Tao et al. (2019) [25] adds a granularity of local information based on the two granularities of DAM, mainly in three forms: Word, Contextual, Attention, and adding word vectors to solve OOV problems. They did a lot of detailed experiments to compare the contribution of the three granularities and the impact of the rounds of conversation and the length of the utterance on the three granularities. The final conclusion is that Contextual contributes the most, and it performs better than Attention when there are few or many rounds.

4   Emotional Analysis

Emotional analysis is a relatively important part. Emotional analysis of the text focuses on analyzing users' emotional situation according to the content and context of the text, and making better responses and processing of the situation according to the results.

Aspect-based sentiment analysis (ABSA) provides more detailed information than general sentiment analysis because its purpose is to predict the emotional polarity of a given aspect or entity in a text. This is done by extracting the relevant aspects, called aspect terms, and detecting the emotional expression of each extracted aspect term, transforming it into an emotion classification at the aspect level. In the past, long-term short-term memory and attention mechanisms have been used to predict the emotional polarity of related objects, which is often complicated and requires more training time.

Neural networks have been widely used in affective analysis and sentence classification. Tree-based recursive neural Tensor Network (Socher et al. (2013) [26]) and Tree LSTM (tree-lstm, Tai et al. (2015) [27]) etc. all carry out syntactic interpretation of sentence structure, but these methods have problems such as low time efficiency and wrong parsing of review text. Recursive neural networks (RNNs) such as LSTM (Hochreiter and Schmidhuber et al. (1997)) and GRU (Chung et al. (2014)) have been used for the analysis of variable length data instances (Tang et al. (2015) [28]). There are also many models using convolutional neural networks (CNNs) (Collobert et al. (2011); Kalchbrenner et al. (2014) [29]; Kim et al. (2014); Conneau et al. (2016) [30]), which also proves that convolution can capture the complex structure of semantically rich text without tedious feature engineering.

Wei Xue, Tao Li et al. (2018) [31] summarized the previous method into two subtasks: aspect categorical emotion analysis (ACSA) and aspect term sentence analysis (ATSA), and proposed a new ACSA and ATSA model, namely gated convolutional network (GCAE) with directional embedding, which is more efficient and simpler than the model based on recursive network (Wang et al. (2016) [32]; Tang et al. (2016); Ma et al. (2017); Chen et al. (2017)) (Figs. 8 and 9).

The model has high precision and efficiency. First, this new gated unit can automatically output emotional characteristics based on a given aspect or entity, which is much simpler than the attention layer used in existing models. Secondly,

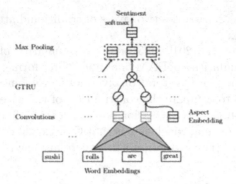

Fig. 8. GCAE for ACSA.

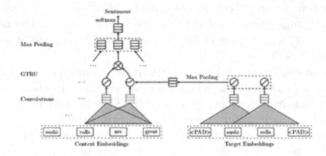

Fig. 9. GCAE for ACSA.

since the convolutional layer is not time-dependent like the LSTM layer, and the gating unit can work independently, the model calculation can be easily parallelized during the training process.

Ruidan He, Wee Sun Lee et al. (2019) [33] proposed an interactive multi-task learning network (IMN) for end-to-end aspect-based emotion analysis, for co-extraction of aspects and viewpoints, as well as emotion classification at the aspect level. It can solve two tasks at the same time, so that the interaction between the two tasks can be better used. In addition, IMN allows AE (aspect term extraction) and AS (aspect level affective classification) to be trained in related document-level tasks, utilizing knowledge from a larger document-level corpus. IMN introduces a new messaging mechanism that allows tasks to interact with each other. Specifically, it sends useful information from different tasks back to the potential Shared representation, and then combines this information with the Shared potential representation to provide all tasks for further processing. This is done iteratively, allowing information to be modified and propagated across multiple links as the number of iterations increases. Among them, a simple way to perform AE and AS simultaneously is multi-task learning, using a Shared network and two task-specific networks to derive a Shared feature space and two task-specific feature Spaces. Multi-task learning adopts the Shared representation, and improves the generalization ability of the model under certain conditions by learning the correlation between tasks in parallel.

The traditional multi-task learning still does not explicitly simulate the interaction between tasks, and the interaction between the two tasks is only to promote learning behavior through false inferences, and this implicit interaction is uncontrollable. IMN not only allows for Shared representations, but also explicitly models interactions.

5 Conclusion

Researches relevant to the field of chatbot conversational systems have been developing rapidly. As the basic work in natural language processing, syntactic analysis has laid a solid foundation for NLP application scenarios such as semantic analysis and emotional expression. Text matching is an essential problem in conversational systems. How to choose a suitable text matching model for different tasks is an important challenge. As people's requirements for chatbots continue to increase, emotional analysis plays an increasingly important role. In this paper, we summarize various existing technologies behind different modules, and compare the advantages and disadvantages of each technology. The modular survey provides a preliminary understanding for beginners who are new to the field of conversational systems, and it is helpful for researchers in related fields to stitch and optimize models.

References

1. Buchholz, S., Marsi, E.: CoNLL-X shared task on multilingual dependency parsing. In: Proceedings of the CoNLL 2006, pp. 149–164 (2006)
2. Chen, D., Manning, C.D.: A fast and accurate dependency parser using neural networks. In: EMNLP, pp. 740–750 (2014)
3. Weiss, D., Alberti, C., Collins, M., Petrov, S.: Structured training for neural network transition-based parsing. In: ACL (1), pp. 323–333 (2015)
4. Ma, M., Huang, L., Zhou, B., Xiang, B.: Dependency-based convolutional neural networks for sentence embedding. In: ACL (2), pp. 174–179 (2015)
5. Kim, Y.: Convolutional neural networks for sentence classification. In: EMNLP, pp. 1746–1751 (2014)
6. Ji, T., Wu, Y., Lan, M.: Graph-based dependency parsing with graph neural networks. In: ACL (1), pp. 2475–2485 (2019)
7. Kunsner, M.J., Sun, Y., Kolkin, N.I., Weinberger, K.Q.: From word embeddings to document distances. In: ICML, pp. 957–966 (2015)
8. Chopra, S., Hadsell, R., LeCun, Y.: Learning a similarity metric discriminatively, with application to face verification. In: CVPR (1), pp. 539–546 (2005)
9. Hu, B., Lu, Z., Li, H., Chen, Q.: Convolutional neural network architectures for matching natural language sentences. In: NIPS, pp. 2042–2050 (2014)
10. Huang, P.S., He, X., Gao, J., et al.: Learning deep structured semantic models for web search using clickthrough data. In: CIKM, pp. 2333–2338 (2013)

11. Shen, Y., He, X., Gao, J., Deng, L., Mesnil, G.: A latent semantic model with convolutional-pooling structure for information retrieval. In: CIKM, pp. 101–110 (2014)
12. Palangi, H., et al.: Semantic modelling with long-short-term memory for information retrieval. CoRR abs/1412.6629 (2014)
13. Qiu, X., Huang, X.: Convolutional neural tensor network architecture for community-based question answering. In: IJCAI, pp. 1305–1311 (2015)
14. Yin, W., Schütze, H.: MultiGranCNN: an architecture for general matching of text chunks on multiple levels of granularity. In: ACL (1), pp. 63–73 (2015)
15. Yu, Z., Liu, G.: Sliced recurrent neural networks. In: COLING 2018, pp. 2953–2964 (2018)
16. Pang, L., Lan, Y., Guo, J., Xu, J., Wan, S., Cheng, X.: Text matching as image recognition. In: AAAI, pp. 2793–2799 (2016)
17. Chen, Q., Zhu, X., Ling, Z.-H., Wei, S., Jiang, H., Inkpen, D.: Enhanced LSTM for natural language inference. In: ACL (1), pp. 1657–1668 (2017)
18. Wang, Z., Hamza, W., Florian, R.: Bilateral multi-perspective matching for natural language sentences. In: IJCAI, pp. 4144–4150 (2017)
19. Yin, W., Schüutze, H., Xiang, B., Zhou, B.: ABCNN: attention-based convolutional neural network for modeling sentence pairs. In: TACL4, pp. 259–272 (2016)
20. Pang, L., Lan, Y., Guo, J., Xu, J., Xu, J., Cheng, X.: DeepRank: a new deep architecture for relevance ranking in information retrieval. In: CIKM, pp. 257–266 (2017)
21. Zhou, X., et al.: Multi-view response selection for human-computer conversation. In: EMNLP, pp. 372–381 (2016)
22. Wu, Y., Wu, W., Xing, C., Zhou, M., Li, Z.: Sequential matching network: a new architecture for multi-turn response selection in retrieval-based chatbots. In: ACL (1), pp. 496–505 (2017)
23. Zhang, Z., Li, J., Zhu, P., Zhao, H., Liu, G.: Modeling multi-turn conversation with deep utterance aggregation. In: COLING, pp. 3740–3752 (2018)
24. Zhou, X., et al.: Multi-turn response selection for chatbots with deep attention matching network. In: ACL (1), pp. 1118–1127 (2018)
25. Tao, C., Wu, W., Xu, C., Hu, W., Zhao, D., Yan, R: Multi-representation fusion network for multi-turn response selection in retrieval-based chatbots. In: WSDM, pp. 267–275 (2019)
26. Socher, R., Perelygin, A., Wu, J.Y., Chuang, J.: Recursive deep models for semantic compositionality over a sentiment Treebank. In: EMNLP, pp. 1631–1642 (2013)
27. Tai, K.S., Socher, R., Manning, C.D.: Improved semantic representations from tree-structured long short-term memory networks. In: ACL, pp. 1556–1566 (2015)
28. Tang, D., Qin, B., Feng, X., Liu, T.: Effective LSTMs for target-dependent sentiment classification. In: COLING, pp. 3298–3307 (2016)
29. Kalchbrenner, N., Grefenstette, E., Blunsom, P.: A convolutional neural network for modelling sentences. In: ACL, pp. 655–665 (2014)
30. Conneau, A., Schwenk, H., Barrault, L., LeCun, Y.: Very deep convolutional networks for text classification. In: EACL, pp. 1107–1116 (2016)
31. Xue, W., Li, T.: Aspect based sentiment analysis with gated convolutional networks. In: ACL (1), pp. 2514–2523 (2018)
32. Wang, B., Liu, K., Zhao, J.: Inner attention based recurrent neural networks for answer selection. In: ACL (1) (2016)
33. He, R., Lee, W.S., Ng, H.T., Dahlmeier, D.: An interactive multi-task learning network for end-to-end aspect-based sentiment analysis. In: ACL (1), pp. 504–515 (2019)

34. Xu, J., Sun, X.: Dependency-based gated recursive neural network for Chinese word segmentation. In: ACL (2) (2016)
35. Rubner, Y., Tomasi, C., Guibas, L.J.: The earth mover's distance as a metric for image retrieval. Int. J. Comput. Vis. **40**(2), 99–121 (2000)
36. Norouzi, M., Fleet, D.J., Salakhutdinov, R.: Hamming distance metric learning. In: NIPS, pp. 1070–1078 (2012)

A Long Short-Term Memory Neural Network Model for Predicting Air Pollution Index Based on Popular Learning

Hong Fang[1(✉)], Yibo Feng[2,1], Lan Zhang[1], Ming Su[1], and Hairong Yang[3]

[1] College of Arts and Sciences, Shanghai Polytechnic University, Shanghai 201209, China
fanghong@sspu.edu.cn
[2] School of Mathematics and Statistics, Kashi University, Kashi 844000, China
[3] School of Mathematics and Statistics, Hefei Normal University, Hefei 230061, China

Abstract. With the acceleration of industrialization and modernization, the problem of air pollution has become more and more prominent, which causing serious impact on people's production and life. Therefore, it is of great practical significance and social value to realize the prediction of air quality index. This paper takes the Tianjin air quality data and meteorological data from 2017 to 2019 as an example. Firstly, random forest interpolation was used to fill in missing values in the data reasonably. Secondly, under the framework of deep learning in TensorFlow, Locally Linear Embedding (LLE) was used to choose multivariate data to reduce data dimensions and realize feature selection. Finally, a prediction model of the air quality index was established by using the Long Short-Term Memory (LSTM) neural network based on the data after dimension reduction. The experimental results show that the method has obvious effects in terms of dimensionality reduction and exponential prediction accuracy compared with Principal Component Analysis (PCA) and Back Propagation (BP).

Keywords: Air quality prediction · LLE · Deep learning · LSTM

1 Introduction

Due to climate change, industrial production and population migration, the air quality in the Beijing-Tianjin wing area is not optimistic. Air quality is closely related to people's health and production and life, haze often hits northern areas. The Air Quality Index (AQI) is an important indicator for people to understand the health of the air, especially PM2.5. The main pollutant which forms smog has become the primary goal of pollution control in Tianjin. Relevant research shows that in addition to pollutant emissions, air quality is also closely related to meteorological conditions such as wind speed, wind direction and temperature etc. Therefore, the analysis of the influence of meteorological changes on the concentration of atmospheric pollutants has important guiding significance for the control and management of pollution sources and the formulation of pollutant treatment plans.

© Springer Nature Switzerland AG 2020
Y. Nah et al. (Eds.): DASFAA 2020 Workshops, LNCS 12115, pp. 190–199, 2020.
https://doi.org/10.1007/978-3-030-59413-8_16

Li Xiaofei et al. [1] analyzed the characteristics of air quality changes in 42 cities in China from 2001 to 2010 and considered that there was a linear relationship between AQI and precipitation and wind speed. Wang liyuan et al. [2] took the air quality of Xichang city as an example, established a variable coefficient model and used the local linear estimation method to fit the model to quantitatively analyze the change of the influence degree of meteorological factors in Xichang city on local air quality with the seasonal change. The prediction based on statistical methods was based on statistics, and the future trend was predicted by analyzing historical air data. With the continuous development of machine learning, it is widely used in the prediction of linear and non-linear and time series models. Common methods mainly include clustering methods [3], support vector machines [4, 5] and neural networks [6] and so on. Among many deep learning models, the recurrent neural network (RNN) introduces the concept of time series into the design of the network structure, which makes it more adaptive in the analysis of time series data. Among many RNN variants, LSTM [7] model solves the problem of gradient disappearance and gradient explosion of RNN, so that the damaged neural network can effectively use long-term time series information.

According to literature review and related study, [8] mainly adopts six AQI indicators such as weather conditions, average temperature, maximum temperature, minimum temperature, wind speed and average wind speed as the initial data research. In view of the non-linear relationship among various indexes which affect the air pollution index, the LLE in the popular learning method is adopted in this paper to reduce the dimension of air data, and the reduced dimension is 8, which are respectively PM2.5, PM10, NO2, O3, weather conditions, minimum temperature, average temperature and average wind speed as input indexes. Based on the characteristics of air quality and meteorological data, an air pollution index prediction model based on LSTM was established in this paper. Compared with the BP neural network model, the LSTM neural network prediction has a lower error rate and it is more suitable for the prediction of air pollution time series data.

2 Algorithm Theory

2.1 LLE Algorithm

Dimension reduction in the field of machine learning refers to a mapping method to map data points in the original high-dimensional space to low-dimensional space. The essence of dimensionality reduction is to learn a mapping function $f : x- > y$, where x is the expression of the original data point, and the vector expression is currently used at most [9]. The common methods include PCA and linear discriminant analysis (LDA), but they are mainly used to deal with linear data sets, which have poor applicability for nonlinear data sets. In 2000, Saul et al. firstly proposed the LLE, which was an unsupervised learning algorithm and can be capable of calculating low-dimensional embedded coordinates of high-dimensional data, and traditional PCA, LDA and other sample-focused variance Compared with the linear dimensionality reduction method, LLE can efficiently maintain the linear characteristics of local samples during dimensionality reduction, and can be widely used in image recognition, high-dimensional data visualization and so on. The main idea of LLE is to keep the linear reconstruction coefficients between the neighbors of high-order data and the linear reconstruction coefficients of the neighbors of low-order data to reduce the dimensionality. Based on this idea, LLE has the characteristics of global nonlinearity and local linearity.

The LLE algorithm is described as follows:

(1) Construct a neighborhood graph. The methods for determining the neighborhood usually include the KNN and ε-nearest neighbor methods. The K nearest neighbor points of each sample point are taken out, and each store is connected to its K nearest neighbor points to form a weighted neighborhood map in high-dimensional data. Solve the local linear matrix between the neighbors of the high-order data. This is achieved by solving the cost function $\Phi(w) = \sum_{i=1}^{N} \left| x_i - \sum_{j=1}^{k} w_{ji} x_{ji} \right|^2$ for getting the minimum w_{ij}. Considering s_i as a local covariance matrix, we can get

$$\begin{cases} S_i = (X_i - N_i)^T (X_i - N_i) \\ \Phi(w) = \sum_{i=1}^{N} w_i^T S_i w_i \end{cases} \tag{1}$$

Using the Lagrangian multiplier method, we can get

$$L(w_i) = w_i^T S_i w_i + \lambda \left(w_i^T 1_k - 1 \right) \tag{2}$$

The value of local reconstruction coefficient can be obtained by differentiating and solving:

$$\begin{cases} \frac{\partial L(w_i)}{\partial w_i} = 2S_i w_i + \lambda 1_k = 0 \\ w_i = \frac{S_i^{-1} 1_k}{1_k^T S_i^{-1} 1_k} \end{cases} \tag{3}$$

(2) Map to a lower dimensional space. In order to optimally maintain the low-dimensional embedded coordinate Y of the weight matrix W, the cost function is:

$$\arg \min_Y \psi(Y) = \sum_{i=1}^{N} \left\| y_i - \sum_{j=1}^{k} w_{ji} y_{ji} \right\|^2 \tag{4}$$

The output is the matrix Y of low dimensional vectors: $Y = [y_1, y_2, \ldots, y_N]$, $d \times N$, W is represented by an N by N coefficient matrix w, for the near point i of j: $W_{ji} = w_{ji}$, so you can get: $\sum_{j=1}^{N} w_{ji} y_{ji} = \sum_{j=1}^{k} w_{ji} y_{ji} = YW_i$. Based on matrix knowledge, $A = [a_1, a_2, \ldots, a_N]$, we can get: $\sum_i (a_i)^2 = \sum_i a_i^T a_i = tr(AA^T)$, use the Lagrangian multiplier method again and derive: $MY^T = \lambda' Y^T$. We can see that Y is actually a matrix formed by the eigenvectors of M. In order to reduce the data to the d dimension, we only need to take the eigenvectors corresponding to the minimum d non-zero eigenvalues of M, and the first smallest eigenvalue is generally close to 0. We discard them and finally keep the feature vector corresponding to the first $[2, d + 1]$ feature values from small to large.

The classic Swiss roll data set in popular learning is used as the experimental data set [10], and the dimensionality reduction using LLE is shown in Fig. 1 to maintain the original topology.

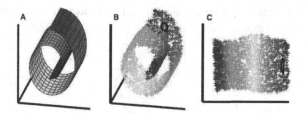

Fig. 1. Schematic diagram of LLE dimension reduction

2.2 Ten-fold cross check

K-fold cross-validation[11] divides the training set into K sub-samples. A single sub-sample is retained as the data of the validation model, while the other $K - 1$ samples are used for training. The cross-validation was repeated K times, once for each subsample, with an average of K results, or using other combinations, resulting in a single estimate. The advantage of this method is that randomly generated subsamples can be repeatedly used for training and verification at the same time, and the results are verified once each time. Figure 2 is the schematic diagram of the ten-fold cross test.

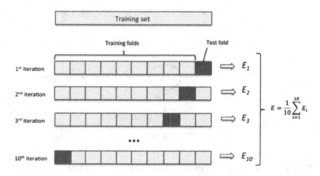

Fig. 2. Schematic diagram of 10-fold cross-validation

2.3 LSTM Algorithm Principle

RNN is a powerful deep neural network, which has been widely used in natural language processing in recent years because of its remarkable effect on time-series data processing with long-term dependence. Training RNN when using back propagation algorithm [12] over time, in order to solve the problems in dealing with long-term dependence on the disappearance of the gradient, Hochreiter & Schmidhuber proposed LSTM model [13], compared with the traditional RNN, it has a more elaborate information transmission mechanism, can effectively solve the problem of long time dependence. At the same time, as the basic subunit in the encoder-decoder framework, LSTM can also realize the encoding and decoding of time series data, and replace the hidden layer neurons in RNN with memory units to realize the memory of past information. Each memory unit contains one or more memory cells and three door controllers. The structure of LSTM is shown in Fig. 3.

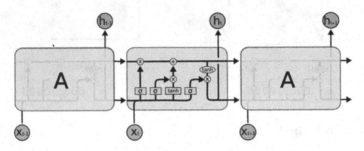

Fig. 3. LSTM structure diagram

The basic structure of LSTM and RNN is similar, with chain structure. However, LSTM has a very useful mechanism, namely forgetting gate. In LSTM, not only a single network layer is used, but four modules are interacted in a special way [14], as shown in Fig. 4.

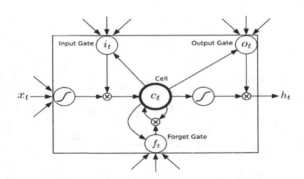

Fig. 4. Schematic diagram of LSTM neural network structure

In the LSTM model, X_t, h_t are respective the input and output of the LSTM network at time t, $t = 1, 2, 3, \cdots$ and the output h_t is the LSTM memory unit, iteratively calculated by the following formula:

$$i_t = \sigma\left(\sum W_{xi}x_t + \sum W_{hi}x_{t-1} + \sum W_{ci}x_{t-1} + b_i\right) \tag{5}$$

$$f_t = \sigma\left(\sum W_{xf}r_t + \sum W_{hy}x_{t-1} + \sum W_{cf}x_{t-1} + b_f\right) \tag{6}$$

$$o_t = \sigma\left(\sum W_{xo}x_t + \sum W_{ho}x_{t-1} + \sum W_{co}x_{t-1} + b_o\right) \tag{7}$$

$$c_t = f_t c_{t-1} + i_t \tanh\left(\sum W_{xc}x_t + \sum W_{hc}x_{t-1} + b_o\right)$$

$$h_t = o_t \tanh(c_t) \tag{8}$$

Among them, i_t, f_t, c_t, o_t represent the vector values of the input gate, forget gate, output gate, and memory unit, w is the weight of various input loops, b is the offset vertex, σ is the sigmoid function, which is used to control the flow of units Weight, ranging from 0 to 1, t is the intercellular activation vector. *Tanh* is a hyperbolic tangent function, and $\odot$ represents an element-wise multiplication operation.

3   Prediction Method of Long-Term and Short-Term Memory Neural Network Based on Popular Learning

3.1   Random Forest Interpolation

Because of the inconsistency between the air quality data and the meteorological data recording time, there are missing values in the acquired data. In order to avoid the impact of missing values on the prediction of the data, choose to use the random forest interpolation method [15] after comparing with other interpolation methods. This method can effectively process high-dimensional data and effectively extract auxiliary variable information, which is suitable for processing issuing data in the context of big data.

3.2   Dimension Reduction

The data used in this article is the meteorological observation data of 30 national meteorological observatories in Tianjin from January 1, 2017 to December 31, 2019. All of the data comes from Tianjin Meteorological Bureau, including weather conditions, wind, wind direction, average wind speed, minimum temperature, maximum temperature and average temperature. The air quality data of Tianjin used in this paper are the daily air quality index of 35 environmental monitoring stations in Tianjin and the concentration data of 6 pollutants (PM2.5, PM10, SO2, CO, NO2, O3). The data comes from Tianjin Environmental Protection Bureau and Tianjin Environmental Protection Monitoring Center.

In this paper, the average classification accuracy before or after dimension reduction is compared by the 10-fold cross-validation technique to verify the effectiveness of the dimension reduction method. The data dimension after dimension reduction is 8 and they are PM2.5, PM10, NO2, O3, weather conditions, minimum temperature, average temperature and average wind speed.

As can be seen from Fig. 5, the accuracy of PCA is higher than that of LLE only when the dimension is reduced to 2. When the dimension is reduced to other number, the accuracy of LLE is significantly better than that of PCA. And the accuracy of the model reached the highest when the dimension is reduced to 8 with an average classification accuracy of 85.07 %. Therefore, LLE is used in the dimensionality reduction of data. This method is effective and the dimension of the data is reduced to 8 with reducing the data volume by about 33 %.

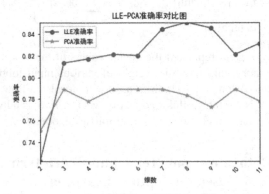

Fig. 5. Accuracy results of LLE and PCA in different dimensions

3.3 LSTM neural network prediction

We choose to use the LSTM neural network and take the first 600 data as the training set, the last 100 data as the test set and then normalize the data. Set the time step, batch size and learning rate for each batch of training samples to 2, 32 and 0.0006 respectively. Then define the neural network variables, use the hyperbolic tangent function (*tanh*) as the activation function in the hidden layer and also use the average absolute error (MAE) loss function and the efficient Adam gradient version of Adam to compile the model. When training this model, the number of iterations (epochs) is 1000 and the comparison between the predicted value (green line) and the real value (red line) is shown in Fig. 6.

3.4 BP neural network prediction

To predict the value of the air quality index AQI, through the previous use of LLE for dimension reduction, 8 factors such as PM2.5, PM10, NO2, weather and wind speed will be used as the impact factors of the API prediction model of the day. Normalizing the data, because of different types of data having different magnitudes and dimensions. Finally, each principal component is used as the input data of the BP neural network model.

The BP neural network prediction model in this paper uses three layers to build, one input layer, one hidden layer, and one output layer. The first layer of activation functions

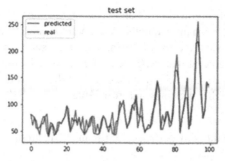

Fig. 6. Prediction Results by LSTM (Color figure online)

uses exponential, and the input variables are two-time steps with 8 characteristics. The second layer uses the softmax activation function for output and the last layer uses the relu activation function for output. Finally, the average absolute error (MAE) loss function and the efficient Adam version of stochastic gradient descent were used to compile the model. The training of the model uses the first 600 data sets as the training set and the last 100 data sets as the prediction set. The image comparison between the final prediction result and the real result is shown in Fig. 7.

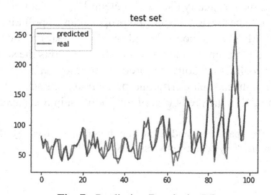

Fig. 7. Prediction Results by BP

To compare LSTM and BP more clearly, we use the indexes of RMSE and Pre-Rate. Test RMSE (standardized) is the root mean square error of the unit data, it represents the model error value of the normalized data with the formula:

$$RMSE = \sqrt{\frac{1}{n} \sum_{i=1}^{n} (y_i - y_i)^2} \tag{9}$$

where y_i is the normalized value and $\hat{y}_i$ is the normalized mean. Test RMSE is the root mean square error of the actual data, it represents the model error value of the real data. Pre_rate is defined by the following formula:

$$Pre\_rate = 1 - \sum_{i=1}^{n} (y_{pre_i} - y_{real_i})^2 \Big/ \sum_{i=1}^{n} (y_{pre_i} - \bar{y}_{real})^2 \tag{10}$$

where y_{pre_i} represents the predicted AQI and y_{real_i} represents the real AQI. Pre_rate represents the similarity between the predicted data and real one. We calculate the predicted value of the root mean square error of normalized data set, root mean square error of real data set and the degree of match between the predicted value and the true value in the prediction results of BP and LSTM, as shown in Table 1, from which we can see that the LSTM model has higher accuracy and robustness.

Table 1. Comparison of BP and LSTM results

Method	Predicted data set test RMSE (Standardized)	Real data set test RMSE	Pre_Rate
BP	0.057	18.284	0.7654394507408142
LSTM	0.047	13.308	0.8757250383496284

4 Conclusion

Aiming at predicting the air quality data and meteorological data of Tianjin, this paper uses LLE method to reduce the dimension, and proposes an air pollution index prediction model based on LSTM. The method is based on a deep learning network, which uses environmental big data to train the model, which fully exploits the semantic features in air quality data, and realizes air pollution prediction based on environmental big data. By comparing the root mean square error and prediction accuracy of LSTM and BP, the validity of LSTM model in the prediction of air pollution in this paper is verified.

Acknowledgments. This paper is funded by the program of the key discipline "Applied Mathematics" of Shanghai Polytechnic University (XXKPY1604).

References

1. Li, X., Zhang, M., Wang, S., Zhao, A., Ma, Q.: Analysis on the change characteristics and influential factors of China's air pollution index. Environ. Sci. **33**(06), 1936–1943 (2012)
2. Wang, L., Li, Y.: Analysis of meteorological factors affecting air quality in Xichang city-a study based on variable coefficient model. J. Yangtze Normal Univ. **33**(05), 98–102+112 (2014)
3. Çevik, H.H., Çunkaş, M.: Short-term load forecasting using fuzzy logic and ANFIS. Neural Comput. Appl. **26**(6), 1355–1367 (2014). https://doi.org/10.1007/s00521-014-1809-4
4. Li, P., Tan, Z., Lili, Y., et al.: Time series prediction of mining subsidence based on a SVM. Int. J. Min. Sci. Technol. **21**(4), 557–562 (2011)
5. Pai, P.F., Hong, W.C.: Forecasting regional electricity load based on recurrent support vector machines with genetic algorithms. Electric Power Syst. Res. **74**(3), 417–425 (2005)
6. Yu, W., Chen, J.: Application of BP artificial neural network model in forecasting urban air pollution. Pollut. Control Technol. **26**(3), 55–57 (2013)

7. Graves, A.: Long short-term memory. In: Graves, A. (ed.) Supervised Sequence Labelling with Recurrent Neural Networks, pp. 37–45. Springer, Berlin (2012). https://doi.org/10.1007/978-3-642-24797-2_4
8. Zheng, Y., Liu, F., Hsieh, H.P.: U-Air: when urban air quality inference meets big data. In: Proceedings of the 19th ACM SIGKDD International Conference on Knowledge Discovery and Data Mining, ACM (2013)
9. Ma, R., Wang, J., Song, Y.: Multi-manifold learning based on nonlinear dimension reduction based on local linear embedding (LLE). J. Tsinghua Univ. (Sci. Technol.) **48**(04), 582–585 (2008)
10. Lan, W., Wang, D., Zhang, S.: Application of a new dimensionality reduction algorithm PCA_LLE in image recognition. J. South Central Univ. Nationalities (Nat. Sci. Ed.) **39**(01), 85–90 (2020)
11. Tang, Y., Xu, Q., Ke, B., Zhao, M., Chai, X.: SVM model optimization of blasting block size based on cross-validation. Blasting **35**(03), 74–79 (2018)
12. Zeng, H.: Research on prediction model of environmental pollution time series based on LSTM. Huazhong University of Science and Technology (2019)
13. Hochreiter, S., Schmidhuber, J.: Long short-term memory. Neural Comput. **9**(8), 1735–1780 (1997)
14. Duan, D., Zhao, Z., Liang, S., Yang, W., Han, Z.: Prediction model of PM2.5 concentration based on LSTM. Comput. Meas. Control **27**(3), 215–219 (2019)
15. Meng, J., Li, C.: Interpolation of missing values of classification data based on random forest model. Stat. Inf. Forum **29**(09), 86–90 (2014)

Domain Ontology Construction for Intelligent Anti-Telephone-Fraud Applications

Shiqi Deng[1], Zhen Zhang[2], and Liang Hong[1($\boxtimes$)]

[1] School of Information Management, Wuhan University, Wuhan, China
1184823582@qq.com, hong@whu.edu.cn
[2] National Computer Network Emergency Response Technical Team/Coordination Center of China, Bejing, China
zhangzhen@cert.org.cn

Abstract. Domain ontology should reflect real data and support intelligent applications. However, existing methods for ontology construction focus on the conceptual architecture while ignore characteristics of domain data and application demands. To enhance the domain applicability of ontology, we propose a method of Data and Application Driven Ontology Construction (*DaDoc*), integrating data characteristics and application demands into the entire construction lifecycle. Our method includes three main phases: data and demands analysis, ontology construction and evaluation. Then, we apply this method to the anti-telephone-fraud field and construct the anti-fraud domain ontology based on cross-domain data. And this ontology represents anti-fraud domain knowledge and it can support intelligent anti-fraud applications, including semantic-based fraud identification and global situation analysis. We finally use quantitative indicators to evaluate the quality of anti-fraud domain ontology. Furthermore, this modeling practice validates the effectiveness of our method for ontology construction and this method can be applied to other domains.

Keywords: Ontology construction · Ontology evaluation · Application-driven · Data-driven · Anti telephone fraud

1 Introduction

Nowadays, the role of domain ontology as the foundation for intelligent applications has become increasingly important. Meanwhile, there is a critical need to enhance the capability of domain ontology to represent real data and support application demands. Ontology is a widely used semantic model with formal specification of shared concepts and machine-understandable format [1]. As the conceptual system extracted from domain data, domain ontology should reflect concepts and features contained in multisource heterogeneous data. Moreover, intelligent applications are supported by ontology in conceptual understanding and knowledge reasoning. Therefore, the design of ontology should consider specific application demands that can vary in different domains. However, existing methods [2–6] for ontology construction focus on concepts and their

© Springer Nature Switzerland AG 2020
Y. Nah et al. (Eds.): DASFAA 2020 Workshops, LNCS 12115, pp. 200–210, 2020.
https://doi.org/10.1007/978-3-030-59413-8_17

architecture, but they ignore the driving force of data and applications. As a result, the quality of ontology oriented to domain applications needs to be improved.

To solve the problems above, we propose a novel method of Data and Application Driven Ontology Construction (*DaDoc*), which integrates data characteristics and application demands into the entire construction lifecycle. We determine the scope of domain data and then analyze data characteristics to extract core concepts of ontology from the bottom up. And we identify domain application demands and then analyze how ontology supports applications to define relationships from top down. Besides, we evaluate whether the built ontology can fully represent domain data and meet application demands. According to evaluation results, the ontology can be optimized to meet with data and application demands with iterations of ontology versions.

In order to validate the modelling method, we take the field of anti telephone fraud ("anti-fraud" for short) as an example and follow *DaDoc* to construct the anti-fraud domain ontology. Anti-fraud is an intersectional domain with complex knowledge associations and demands for intelligent anti-fraud applications are emerging faced with complicated and varied fraud patterns [6]. Therefore, the construction of anti-fraud domain ontology should be driven by data and applications. Using *DaDoc,* we extract fraud content features from real call transcript and use semantic information to expand analysis models that only reflect fraud behavior features [8–12]. We establish a semantically rich view of global anti-fraud knowledge from main aspects of fraud patterns, business permissions and fraud gangs. And the ontology constructed is capable to provide knowledge association and global view for fraud telephone identification and fraud situation analysis. This modeling practice also validate the effectiveness of our method for ontology construction.

2 Related Work

2.1 Domain Ontology Construction and Evaluation

Representative ontology construction methods include Seven-Step method [2], Skeleton method [3], Circulation method [4], IEEF5 [5], Meth-Ontology [6] etc. The first two methods have high maturity and are widely used by scholars in various fields. Seven-Step method defines specific steps of building ontology with feasible instructions but lacks ontology evaluation. Skeleton method provides the whole process of building ontology including ontology evaluation but lacks detailed description of steps. Ontology evaluation is an important part for optimizing ontology logical structure and function. Xu [13] reviewed existing major methods and tools for ontology evaluation and the multi-criteria method based on various evaluation indicators gained reputation for its comprehensiveness and quantitative observation. Hloman [14] explored multiple dimensions of data-driven ontology evaluation and measured the impact of domain knowledge dimensions on ontology evaluation by time bias and category bias.

In recent years, ontology construction has been initially explored about telecommunication fraud. In 2017, Yang et al. [8] established a knowledge framework for telecommunication fraud analysis based on ontology and they divided the fraud implementation into four elements as information stealing, deception concealment, means implementation and withdrawal. However, this ontology is coarse-grained without abundant semantic relationships. There are many ontology researches in related fields such as network security and financial fraud detection. Park et al. [15] proposed the Cyber Forensics Ontology according to case categories, laws, evidence and criminals, and they improved the efficiency of investigation by data mining. Carvalho et al. [16] constructed an ontology to identify malware invading the banking system and map criminal organizations. Fauzan et al. [17] offered an ontology matching approach based on the structure of the business process ontology to detect business fraud.

2.2 Anti-fraud Method

Becker et al. [7] sorted out the history of fraud detection algorithms as follows: early threshold-based alerting, signature-based alerting, moving to graph-based signatures and catching fraud via graph matching. It is observed that anti-fraud methods in telecommunications are mainly based on various probability statistical models such as the Latent Dirichlet Allocation [9] and Gaussian Mixture Mode [10]. In addition, common anti-fraud models [8, 11] are based on fraud call behaviors extracted and trained from Call Detail Records (CDRs) which are much easier to obtained because signaling acquisition systems are increasingly improved. Then data mining algorithms and machine learning are applied to analyze and intercept telephone numbers.

Except for CDRs mining and analysis, telephone speech recognition is also an important technology in the anti-fraud field. Ma [12] established a fraudulent speech waveform model and built an anti-fraud system based on speech content analysis. But it still relies on static collections of sensitive words that are artificially constructed to match suspected telephone numbers with existing blacklists. It can be seen that existing anti-fraud schemes mainly analyze the characteristics of call behaviors based on CDRs and voice data. However, none of these models take full advantage of semantic association contained in fraud call contents, which makes it difficult to fully represent and identify fraud patterns, let alone grasp overall fraud situation.

3 Methodology

3.1 Overview

Our general strategy for ontology construction is to incorporate data-driven and application-driven approaches to make data and application demands play a role in the entire lifecycle of ontology construction. Consequently, the domain ontology constructed is more capable of representing domain knowledge and supporting intelligent applications. With the reference to both Seven-Step method and Skeleton method, our method *DaDoc* provides not only the whole process of building ontology including evaluation, but it also refines specific steps considering the impact of data and applications.

As shown in Fig. 1, *DaDoc* is planned in three phases: (1) data and demands analysis: an adaptation on the traditional step of determining the domain scope of ontology, to make sure before concrete design that ontology is based on data and supportive of applications. In terms of data, we focus on data sources and their accessibility, data format and modeling features, and data fields involved in contents. In terms of application, we find limitations of current applications and refine unsatisfied domain demands, then design the implementation path at ontology layer; (2) ontology construction: an integration of considering ontology reuse, defining classes and facets, creating attributes and relationships, and each step is well-founded from data and demands; (3) ontology evaluation: combining both structure and function dimensions to assess whether the ontology can fully represent domain data and whether its logical association can meet application demands, if not, return to the ontology construction phase and continuously optimize the ontology after several iterations of ontology version until it meets with data characteristics and application demands.

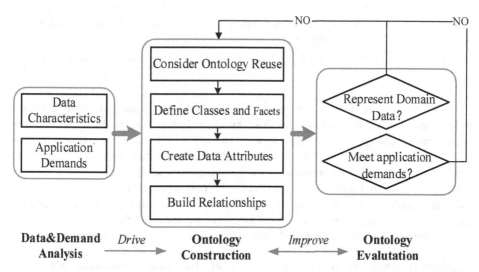

Fig. 1. *DaDoc*: Data and Application Domain Ontology Construction Method.

Then, we will apply the domain ontology construction method to the anti-fraud filed, thus making the anti-fraud domain ontology proposed in accordance with anti-fraud data characteristics and intelligent anti-fraud application demands. We extract fraud content features from real call transcript and use such semantic information to expand analysis models that only reflect fraud behavior features. The ontology constructed is capable to provide knowledge association and global view for fraud telephone identification and fraud situation analysis. Important concepts and their relationships related to the anti-fraud field used in this paper are summarized in Table 1.

3.2 Data and Demands Analysis

Data Characteristics. Reverse data is the core basis for fraud telephone identification, from which fraud means and processes can be analyzed. (1) Fraud event data: mainly from the news about telephone fraud on the Internet. (2) Fraud call data: nationwide records of fraud calls obtained from this emergency management department [18] in the past five years. It can be divided into two types as CDRs and transcript of fraud calls, which record call behaviors and contents respectively. The latter is text data parsed by speech recognition technology and records the entire fraud process.

Table 1. Definition of important concepts in the anti-fraud field.

Concepts	Explanation
Fraud event	Short name of telephone fraud event, including people, call behaviors and contents, fraud processes. Our ontology models each part of fraud event
Fraud scene	Background information used by fraudsters, which is the main basis for distinguishing the types of fraud events
Fraud process	Process of fraudsters committing fraud composed of sequential steps, which is the reflection of fraud means and the core of fraud analysis
Reverse data	Telephone fraud events and their call records. The information provided by fraudsters during calls is false and fraudulent
Forward data	Correct and normative information related to fraud scenes in various fields

However, reverse data is lagging historical records and make anti-fraud decisions and actions confused by fraudsters and lose initiative. To solve this problem, we find relevant forward data and then correlate forward and reverse data to actively recognize and early resolve vulnerabilities that are easily used. Forward data is mainly derived from regulations on related institutions and their business on official websites. As is showed in the Chinese official statistical report [19], the proportion of such fraud scenes as financial institutions, e-commerce platforms, public security organs and telecom operators accounted for more than 67% in 2017. Therefore, we focus on the above four typical types of fraud scenes and will flexibly adjust to the changes of fraud scenes.

Application Demands. Fraud telephone identification based on semantic understanding: mine and understand more semantic information such as fraud processes in calls, so as to inspect semantic inconsistency of call behaviors and contents and determine the fraud probability and fraud means of any single fraud telephone. To this end, the anti-fraud domain ontology is required to fully represent semantic association of anti-fraud field, that is, to conduct reverse modeling for fraud events and forward modeling for business permissions related to fraud scenes, and then make them semantically associated and compared.

Fraud situation analysis with a global view: analyze characteristics, relevance and evolution of fraud events in multi-dimensions including regions, time, fraud means, fraud gangs and vulnerable groups, supporting upper anti-fraud decisions. To this end, the anti-fraud domain ontology is required to model overall cross-domain knowledge of anti-fraud field and the emphasis of this application is to portray fraud gangs, which can be used to correlate each feature of fraud events. Besides, recognizing gangs to which fraudsters belong from suspected fraud calls can further support semantic-based fraud telephone identification.

Finally, the framework of anti-fraud domain ontology construction driven by data and application is designed in Fig. 2. It shows the association among three layers as data, ontology and application. Accordingly, ontology construction in the following phase are divided into three modeling parts of reverse fraud event, forward business permission and fraud gang portrait.

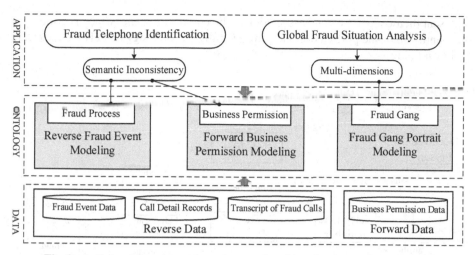

Fig. 2. Anti-fraud domain ontology construction driven by data and application.

3.3   Ontology Construction

Reverse Fraud Event Ontology. Reverse modeling is the main part of the anti-fraud modeling scheme. It is designed to analyze and represent the means and processes of telephone fraud, according to which, machine can compare the similarity between suspicious call contents and fraud calls to identify fraud telephones. We summarize a general fraud process from real fraud texts and then regulate it in the form of ontology to integrate it into the anti-fraud model, laying the foundation for semantic-based fraud telephone identification. Reverse modeling mainly focuses on such three classes as "Fraudster", "Deceived Person" and "Fraud Process". The main relationship "call" can associate two types of participants, because fraudsters tend to call up deceived people actively in most of real fraud events.

Extracting Common Fraud Pattern. We analyze a large amount of transcript of fraud calls and sum up a common three-step fraud process: (1) the first step is to gain the trust of others and fraudsters will fake an identity; (2) intermediate steps is to induce others to fall into the pre-designed fraud trap and fraudsters usually make up causes and request designated operations, which involve permissions of the business entity faked by fraudsters; (3) the last step is the key for fraudsters to defraud money and complete a fraud, where they trick others to transfer directly to appointed accounts or defraud private information like credit card numbers and passwords.

Modeling of Fraud Process. With referenceto the above general fraud pattern, we decompose the class "Fraud Process" into three subclasses as "Faking Identity", "Inducing Process" and "Defrauding Money", and the relationship among them is "next step". In particular, "Fraud Process" has two types of single scene and composite scene: the former refers to fraud process that only occurs in a set scene, and in the latter type there is an operation of transfer to another specified number, thus jumping to another fraud process. To this end, we design the relationship "transfer" between fraudsters to represent their cooperation and association in the composite scene, which is critical for the identification and portrait of fraud gangs. Finally, the ontology model of reverse data is shown in Fig. 3.

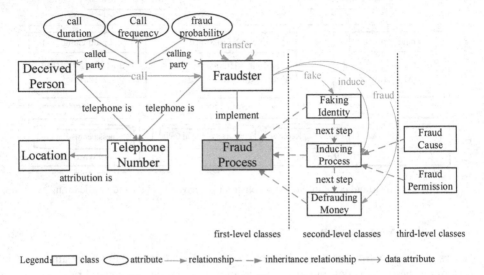

Fig. 3. Ontology model of inverse fraud event.

Forward Business Permission Ontology. Forward modeling is the new perspective of anti-fraud scheme pioneered in this paper. It aims to analyze and represent the business entities and their permissions related to fraud scenes. Telephone fraud uses the false information of business entities, which violates normal business permissions. Anti-fraud only from the reverse struggles with the change of fraud means. On the contrary, grasping in advance the normative connotation of business permissions and their parts that are easy to be used will make the anti-fraud scheme more proactive. Besides, the comparison between the forward and reverse model can assist in fraud identification. Accordingly, we design two classes as "Scene Entity" and "Business Permission", and there is a relationship as "have permission as".

Modeling of Scene Entity. This class shows business subjects corresponding to telephone fraud scenes and it is associated with "Faking Identity" which is the first step of "Fraud Process" through the relationship as "correspond to subject". At present, we divide "Scene Entity" into five common subclasses and they can be flexibly expanded based on actual fraud scenes.

Modeling of Business Permission. This class represents the normal business contents and related regulations of scene entities, with attributes involving category, content, basis, material required, time and approach. These attributes are often used by fraudsters to design fraud processes, so we can semantically compare them to the similar parts mentioned in fraud processes, thus detecting inconsistency in calls. Finally, the ontology model of forward data is shown in Fig. 4.

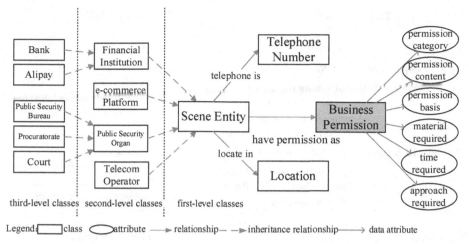

Fig. 4. Ontology model of forward business permission.

Fraud Gang Portrait Ontology. Fraud gangs are a kind of criminal organization with similar fraud patterns and internal cooperation. The key to identifying and monitoring fraud gangs is to construct fraud gang portraits based on fraud patterns accurately and comprehensively. Portraits mainly describe the class "Fraud Gang" and involve the refinement of other classes like "Telephone number" and "Fraud Process". And the class "Fraudster" corresponds to "Fraud Gang" through the relationship as "belong to gang". We analyze and model the fine-grained fraud means from transcript of fraud calls and then combine two major groups of attributes as "number features" and "means features" to describe call methods and contents of fraud gangs.

Modeling of Number Features. This group summarizes attributes of the class "Telephone Number". The class "Fraud Gang" has three special number attributes to identify fraud gangs and assist in anti-fraud: (1) number behavior features describe historical call behaviors of fraud gangs; (2) number activeness features describe the status throughout the life cycle of fraud numbers; (3) number association features contribute to sort out social networks between fraud numbers.

Modeling of Means Features. This group can be interpreted as various kinds of "Fraud Process" that is commonly used by all the "Fraudsters" of each "Fraud Gang". It can be divided into five attributes that respectively list the fake identity, inducement, permission used, transfer range and way to defraud money of specific fraud gangs by frequency.

3.4 Ontology Evaluation

Evaluation Methods and Indicators. We apply a multi-criteria ontology assessment tool *OntoQA* [20] to quantitative evaluation of ontology architecture and mainly use Schema Metrics whose specific indicators are shown in Table 2. Besides, we compare our anti-fraud domain ontology (hereinafter "Anti-O") with the telecommunication fraud analysis ontology (hereinafter "Tele-O") [8] constructed by Yang et al.

Table 2. Definition of important concepts in the anti-fraud field.

Indicator	Formula	Meaning
Relationship Richness	(1) $RR = (|P|)/(|SC| + |P|)$	Indicates the diversity of relationships. An ontology with more other relationships tends to express more information
	|SC| is the number of inheritance relationships; |P| is the number of other relationships except inheritance relationships	
Attribute Richness	(2) $AR = (|att|)/(|C|)$	Indicates the diversity of attributes. An ontology with more attributes tends to be richer in connotations and have higher quality
	|att| is the number of attributes for all classes; |C| is the total number of classes	
Inheritance Richness	(3) $IR_C = \left(\sum_{C_i \in C} \left| H^C(C_1, C_i) \right| \right)/(|C|)$	Describes the number of inheritance relationships at different levels of ontology and draws the structure form of ontology
	$\left| H^C(C_1, C_i) \right|$ is the number of subclasses (C1) for a class (Ci)	

Evaluation Results and Analysis. After the quantitative statistics of classes, attributes and relationships, the indicators of are shown in Table 3. (1) Main relationships of Tele-O is of inheritance, while other types of relationships are less, mostly "attribute-of", and the specific quantity is unknown. Whereas, Anti-O customizes more types of other relationships based on actual field conditions and its Relationship Richness 0.797 is close to 1, which indicates that Anti-O has more diverse relationships and it is more effective in fraud identification through knowledge association. (2) Attribute Richness will not be horizontally compared because this indicator of Tele-O is unknown. Anti-O has an Attribute Richness as 4.148, which is at a relatively high level and indicates that Anti-O is capable of representing anti-fraud domain knowledge comprehensively. (3) Anti-O has smaller Inheritance Richness than that of Tele-O, so the structural form of Anti-O is more horizontal with a large number of inheritance levels where classes have a small number of subclasses are vertical. This result means that Anti-O represents a wide range of general knowledge and it is consistent with the complex nature of anti-fraud knowledge.

In addition to ontology structure, we also consider the evaluation dimension of ontology function. Our application-driven ontology construction process can ensure that the anti-fraud domain ontology is satisfied with the application requirements to a certain extent. Specifically, based on the semantic understanding of ontology, the similarity between suspicious processes in calls and typical fraud processes refers to fraud probability, and the differences between suspicious steps in processes and normal business permissions related to fraud scenes can assist in fraud identification. Besides, using the global view provided by ontology, we can understand high-incidence area and

Table 3. Evaluation result comparison of Anti-O and Tele-O.

Indicators	Anti-O	Tele-O
Number of inheritance relationships	27	53
Number of other relationships	106	Unknown
Relationship Richness	**0.797**	**Unknown**
Number of all classes	61	54
Number of attributes for all classes	253	Unknown
Attribute Richness	**4.148**	**Unknown**
number of subclasses for all classes	114	139
Inheritance Richness	**1.869**	**2.574**

peak period of fraud calls, geographical distribution and high-frequency fraud means of fraud gangs, and also dynamic evolution on all dimensions of fraud events.

4 Conclusions

We propose a domain ontology construction method for intelligent applications, which integrate data characteristics and application demands into entire traditional process of Seven-Step methodology and Skeleton methodology. Then we apply this method to construct the anti-fraud domain ontology and main features of this ontology such as the detailed modeling of fraud process, the association of forward and reverse data and the comprehensive portrait of fraud gang contribute to intelligent fraud identification and situation analysis. The ontology evaluation based on quantitative indicators shows that this ontology constructed can represent anti-fraud knowledge and this method proposed can make ontology in accordance with data and application demands.

Acknowledgement. This paper is supported by National Key Research and Development Project No. 2018YFC0806900.

References

1. Gruber, T.R.: A translation approach to portable ontology specifications. Knowl. Acquisition **5**(2), 199–220 (1993)
2. Noy, N.F., Mcguinness, D.L.: Ontology development 101: a guide to creating your first ontology. Stanford Knowledge Systems Laboratory (2001)
3. Mike, U., Michael, G.: Ontologies: principles, methods and applications. Knowl. Eng. Rev. **11**(2), 93–136 (1996)
4. Maedche, A., Staab, S., et al.: Ontology learning for the semantic web. Intell. Syst. IEEE **16**(2), 72–79 (2001)

5. Ye, Y., Yang, D., Jiang, Z., et al.: Ontology-based semantic models for supply chain management. Int. J. Adv. Manuf. Technol. **37**(11–12), 1250–1260 (2008)
6. Hengjie, L.I., Junquan, L.I., Ming, L.I.: Research on domain ontology modeling method. Comput. Eng. Des. **29**(2), 381–384 (2008)
7. Becker, R.A., Volinsky, C., Wilks, A.R.: Fraud detection in telecommunications: history and lessons learned. Technometrics **52**(1), 20–33 (2010)
8. Yang, J., Cao, J., Gao, H.: Telecommunications fraud case analysis knowledge base model based on ontology. Comput. Eng. Des. **38**(06), 1418–1423 (2017)
9. Olszewski, D.: A probabilistic approach to fraud detection in telecommunications. Knowl.-Based Syst. **26**, 246–258 (2012)
10. Yusoff, M.I.M., Mohamed, I., Bakar, M.R.A.: Improved expectation maximization algorithm for Gaussian mixed model using the kernel method. Math. Probl. Eng. **1**, 377–384 (2013)
11. Li, L., Ma, Z., Chen, Q., Li, C.: Research of technology solutions and operation counter-measures to telephone fraud prevention and control. Telecommun. Sci. **30**(11), 166–172 (2014)
12. Ma, B.: A method and system for anti-telephone fraud based on speech semantic content analysis: CN, CN 103179122 B[P] (2015)
13. Lei, X.: Progress in ontology evaluation. J. China Soc. Sci. Tech. Inf. **35**(07), 772–784 (2016)
14. Hloman, H., Stacey, D.A.: Multiple dimensions to data-driven ontology evaluation. In: International Joint Conference on Knowledge Discovery, Knowledge Engineering, and Knowledge Management, pp. 329–346 (2014)
15. Park, H., Cho, S., Kwon, H.-C.: Cyber forensics ontology for cyber criminal investigation. In: Sorell, M. (ed.) e-Forensics 2009. LNICST, vol. 8, pp. 160–165. Springer, Heidelberg (2009). https://doi.org/10.1007/978-3-642-02312-5_18
16. Carvalho, R., Goldsmith, M., Creese, S.: Applying semantic technologies to fight online banking fraud. In: Intelligence and Security Informatics Conference, pp. 61–68. IEEE (2015)
17. Fauzan, A.C., Sarno, R., Ariyani, N.F.: Structure-based ontology matching of business process model for fraud detection. In: International Conference on Information & Communication Technology and System, pp. 221–226 (2017)
18. National Internet Emergency Center[EB/OL]. http://www.cert.org.cn/
19. Telephone State, Trusted Number Data Center. Harassment, fraud telephone situation analysis report in [EB/OL] (2017). https://www.kexinhaoma.org/Page/News
20. Tartir, S., Arpinar, I.B., Moore, M., et al.: OntoQA: metric-based ontology quality analysis. In: IEEE ICDM 2005 Workshop on Knowledge Acquisition from Distributed, Autonomous, Semantically Heterogeneous Data and Knowledge Sources. IEEE (2005)

Improving Gaussian Embedding for Extracting Local Semantic Connectivity in Networks

Chu Zheng[✉], Peiyun Wu, and Xiaowang Zhang

College of Intelligence and Computing, Tianjin University, Tianjin 300350, China
zhengchu@tju.edu.cn

Abstract. Gaussian embedding in unsupervised graph representation learning aims to embed vertices into Gaussian distributions. Downstream tasks such as link prediction and node classification can be efficiently computed on Gaussian distributions. Existing Gaussian embedding methods depending on adjacent neighbors of vertexes leave out of consideration of indirect connectivity information carried by remote vertice, which is still a part of local semantic connectivity in a network. In this paper, we propose an unsupervised graph representation model GLP2Gauss to improve Gaussian embedding by coarsening graph as indirect local connectivity. First, we decompose the original graph into paths and subgraphs. Secondly, we present a path-based embedding strategy combined with the Gaussian embedding method to maintain better local relevance of embedded nodes. Experiments evaluated on benchmark datasets show that GLP2Gauss has competitive performance on node classification and link prediction for networks compared with off-the-shelf network representation models.

Keywords: Social network · Graph representation learning · Gaussian embedding

1 Introduction

Network representation learning (NRL) is to embed all vertices in graphs into low-dimensional dense vectors, which is essential to network analysis. A big challenge of NRL is about keeping nearly original information of input graphs into low-dimensional embeddings, and also getting good properties for the output specific learning tasks.

As an important class of NRL approaches, unsupervised graph representation learning [20] extracts positive and negative samples from the graph by a scoring function to give high scores to positive samples and low scores to negative samples. The key of this method is how to define the scoring function (or proximity function) [7] to further rank training samples. Several unsupervised graph representation approaches have been proposed, such as random-walk based and Gaussian embedding approaches.

© Springer Nature Switzerland AG 2020
Y. Nah et al. (Eds.): DASFAA 2020 Workshops, LNCS 12115, pp. 211–224, 2020.
https://doi.org/10.1007/978-3-030-59413-8_18

Random-walk based methods focus on direct and indirect connectivity in graph [11]. Most existing random-walk based approaches [8] obtain a subgraph around each vertex as a positive sample, and assign corresponding accessible probabilities to the vertices in the subgraph. Thus connections with adjacent and non-adjacent vertices in the subgraph are established. For instance, the DeepWalk model [2] adopts a random sampling (walk) strategy for each vertex in networks to generate a bunch of vertex sequences as context vertices. The DeepWalk model can maximize the posterior probability of each node's context vertices. Since then, similar algorithms have been proposed, such as node2vec [3], TriDNR [4] etc. The success of these models in node classification, link prediction, and other tasks, illustrating the importance of direct and indirect connectivity information.

Recently, many graph representation methods focus on Gaussian embedding [5,6]. That is, each vertex is embedded into a Gaussian distribution; we illustrate an example in Fig. 1. The first Gaussian embedding approach [12] is proposed to learn the uncertainty of word embeddings. In the Knowledge Graph, another Gaussian embedding approach [13] computes the (un)certainties of entities and relations, and [5] employs asymmetric KL divergence on Gaussian embedding in directed attribute graph. [6] applies 2nd Wasserstein distance to optimize node embeddings in the undirected graph. Compared to random-walk based methods that treat vertices as conventionally latent vectors, embed vertices as Gaussian distributions allow to capture the uncertainty of vertex representations [5].

For example, the embedding of low degree vertices generally contain more uncertainty since they have less connected vertices constraints and the high degree vertices are opposite.

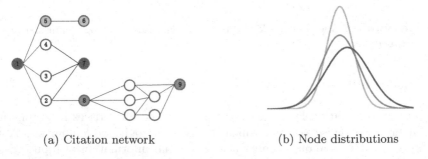

(a) Citation network   (b) Node distributions

Fig. 1. (a) is a part of the paper citation network, where each node is a paper in the network, and each edge represents citation relationships. Suppose the same colored node pairs are assigned high accessible probabilities to each other by random-walk based models. Thus their embeddings can have highly similar semantic characteristics. For Gaussian embedding models, similar vertices pairs are embedded into similar Gaussian distributions (b). (Color figure online)

Both random-walk based and Gaussian embedding approaches are contrastive methods [21]. However, existing Gaussian embedding models mainly focus on adjacent relations (direct connectivity). They extract connected vertex pairs as positive samples and non-connected vertex pairs as negative samples. Therefore, indirect connectivity information is ignored. There are two main reasons for this. The first is that existing Gaussian embedding models use contrastive loss function, which may lead model performance to suffer from poor local optima [24]. The second is that, in unsupervised graph learning, node pairs coming from similar distributions (especially non-connected node pairs) are difficultly defined. Thus, existing Gaussian embedding models [5,6] mainly consider 1-hop connected node information by treating edges in the network as equal. As a result, node embeddings are sub-optimal.

In this paper, we propose a novel Gaussian embedding model GLP2Gauss (Graph Local Path substructure) by considering both direct connectivity and indirect connectivity in unsupervised graph learning. We present a new novel structure including *Path* and *Subgraph* in a network, all indirectly connected vertices in *Path* can be embedded into similar Gaussian distributions. We prove that vertices in *Path* have the property of a strong correlation. Moreover, we present a path embedding method combined with Gaussian embedding to embed the path node into a similar Gaussian distribution. Finally, several experimental tasks show that GLP2Gauss outperforms baselines and achieves competitive results on benchmarks.

2   Related Works

The main popular approaches proposed for NRL can be classified into two categories: *unsupervised learning* and *semi-supervised learning*.

Unsupervised Learning. Inspired by the Skip-Gram [1], DeepWalk [2] treats vertices as *words* and node sequences based on random-walks as *sentences*, learning embeddings via the prediction of posterior probability between nodes and its neighboring nodes, and similar algorithms have been proposed [3,4]. Besides, [5,6] assumes that the node embedding obeys Gaussian distribution and uses the node pairs ranking strategy to learn node embedding. Different from the above methods based on local similarity, DGI [21] learns global patch representation via contrast global and local mutual information of embeddings. In summary, this class of methods assume local or global similarity in the graph.

Semi-supervised Learning. Another kind of approach is graph neural networks [20,22,23], which extends the convolutional neuron network from euclidean to non-euclidean data. These methods follow the message-passing architecture and focus on using a small number of labeled nodes for end-to-end node or graph classification tasks.

3   Overview of GLP2Gauss

In this section, we briefly introduce the problem statement and the overview of our proposed GLP2Gauss model.

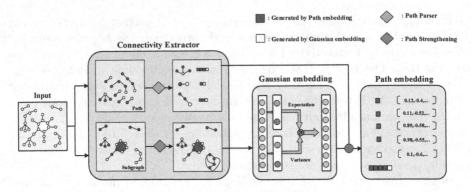

Fig. 2. The framework of GLP2Gauss

Suppose we have a network $G = (V, E)$ (an undirected and no attributes graph) where V is a set of vertices and E is a set of edges. The goal of NRL is to learn a mapping $f : R^{|V| \times |V|} \to R^{|V| \times |V|}$ by minimizing the loss functions of models. Formally, the goal can be formalized as follows: $\min_Y \mathcal{L}(f(A), g(Y))$, where $Y \in R^{|V| \times d}$ is a d-dimensional embedding matrix, $A \in \{0, 1\}^{|V| \times |V|}$ is the adjacent matrix of graph G, $g : R^{|V| \times d} \to R^{|V| \times |V|}$ is a transformation function, $\mathcal{L} : R^{|V| \times |V|} \to R^{|V| \times |V|}$ is a loss function. In Gaussian embedding models [5], each node in a network is embedded into a vector in the Gaussian distribution (or normal distribution). Let $Y = [y_i]_{|V|}$ with $y_i = \mathcal{N}(\mu_i, \Sigma_i)$, $\mu_i \in R^d$, and $\Sigma_i \in R^{d \times d}$. In this case, g is taken as a dissimilarity function between two distributions. Inspired by random-walk based models [7] strengthening the relatedness between vertices in local scope, we integrate partial path information into Gaussian embedding model to enrich node embeddings.

The framework of GLP2Gauss consists of three modules, namely, *Connectivity Extractor*, *Gaussian embedding*, and *Path embedding* (shown in Fig. 2) and the three modules of GLP2Gauss model are as follows:

- Connectivity extractor decomposes the original input graph into a set of paths and a set of subgraphs. Within this module, we present two submodules: *Path Coarsening* and *Path strengthening* for the original graph decomposition. As shown in Fig. 2, the output path set is used as the input of the Gaussian embedding module, and the output subgraph set is used as the input of the path embedding module.
- Gaussian embedding is to learn embeddings of the subgraphs, which is generated in the Connectivity extractor. By Gaussian embedding, GLP2Gauss can obtain embedding of those nodes on subgraphs.
- Path embedding is to generate embedding of nodes on paths. The path embedding module follows the Gaussian embedding module. In this way, we can obtain the full embedding of nodes on the original graph.

4   GLP2Gauss: An Embedding NRL Model

In this section, we present the GLP2Gauss learning strategy to learning Gaussian embeddings of nodes.

4.1   Connectivity Extracting via Paths

We use P to denote a path set whose middle vertices have exactly 2 degrees in an undirected graph, and the endpoints of the path are represented by s and t. We present two approaches to connectivity as follows:

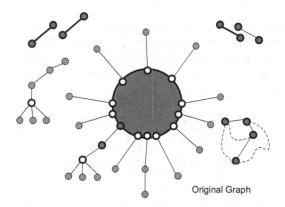

Original Graph

Fig. 3. The red nodes represent the subgraph of the original network, and the blue nodes represent the coarsening paths. The white vertices remain on these paths. Moreover, the two vertices connected by dot line means that they are embedded into similar Gaussian distributions via path strengthening. (Color figure online)

Path Coarsening. First, we use the connectivity extractor to extract the coarsening path set P_C from original graph G_O, then the original graph G_O is derived from summarizing G_C via path set P_C. The coarsening path set P_C is a subset of P, which satisfies $p_{st} \in P_C, |D(s)| = 1 \wedge |D(t)| \neq 1$ or $|D(s)| \neq 1 \wedge |D(t)| = 1$ where $|D(s)|$ and $|D(t)|$ is the degree of node s and t. For each path p_{st} in P_C, if $|D(s)| \neq 1$ (for $|D(t)| \neq 1$ is similar to this), the path coarsening operation makes $s \longleftarrow s+1, s+2, \cdots t$, that is, path p_{st} is pruned directly from the original graph, leaving only node s on the original graph, so we have subgraphs as follows: $G_C \longleftarrow G_O, v_s \longleftarrow p_{st}$, where G_C is a set of coarsening subgraphs of the original graph G_O. In this way, for each path P_C, we coarsen it as a node, and obtain coarsening graph in G_C, an example is illustrated in Fig. 3, we coarsen the original graph into dense subgraphs (colored by red).

Path Strengthening. For paths $\tilde{p}_{st} \notin P_C$ and $\tilde{p}_{st} \in P$, we connect all vertices of each $\tilde{p}_{st}$ to strengthen them into similar Gaussian distributions. For each node i and j in path $\tilde{p}_{st}$, if node i is not connected with j, we create a pseudo edge between i and j, for example, in Fig. 3, the satisfied path at the bottom right is strengthened by created edges between vertices in path, note that we also strengthen previous defined path that are inside of the coarsening subgraph.

4.2   Gaussian Embedding

In this section, we introduce the Gaussian model for training coarsening subgraph set G_C.

Model Structure. We use the [5] proposed model structure to perform Gaussian embedding on G_C. Firstly, the input x_i are mapped to the encoding layer, which contains 512 hidden neurons and gets hidden embedding y_i. Secondly, the results y_i are mapped to μ and σ hidden layers with 64 dimensions. Then we have μ layer and σ layer as follows:

$$
\begin{aligned}
y_i &= \mathbf{Relu}(x_i W^{(1)} + b^{(1)}) \\
\mu_i &= y_i W^{(2)} + b^{(2)} \\
\sigma_i &= \mathbf{Elu}(y_i W^{(3)} + b^{(3)}) + 1 \\
Y_i &= \mu_i \oplus \sigma_i
\end{aligned}
\tag{1}
$$

where **Relu** and **ELU** are activation functions. Finally, together μ and σ to form the node Gaussian embedding Y, note that $\oplus$ means a concatenation operation between μ and σ.

Transformation Function. As we embed all vertices into Gaussian distribution, we follow [6] and choose 2th Wasserstein distance as function g, which is suitable for undirected graphs. We assume the covariance matrices of node embedding are diagonal. Then the 2th Wasserstein distance has the closed-form solution [14] belong:

$$
g_{(v_1, v_2)} = \|\mu_1 - \mu_2\|_2^2 + \mathrm{Tr}(\Sigma_1 + \Sigma_2 - 2(\Sigma_1^{1/2} \Sigma_2 \Sigma_1^{1/2})^{1/2})
\tag{2}
$$

The above equation embeds the latent node embeddings to be closer if node v_1 shares an edge with node v_2, on the contrary, for each positive node pairs we sample corresponding negative node pairs with non-edges in graph and push them away.

Loss Function. In order to rank the true edges more closer than pseudo edges, we choose an energy-based function [9] to calculate all loss of node pairs:

$$
L_2 = \sum_{i,j,k \in V} (g(V_{ij})^2 + exp(-g(V_{ik})))
\tag{3}
$$

where V_{ij} and V_{ik} are positive and negative samples, respectively. Notice that node pairs obtain from paths are all treated as positive samples. Furthermore, we make a $\mathcal{L}1$ loss for each node embedding Y_i minus its corresponding 1-hop neighbour node embedding Y_j:

$$\mathcal{L}_1 = \|Y_i - Y_j\|_1$$

Since our embedding follows Gaussian distribution, we only make the loss on the μ part of each node embedding. Thus we have the final loss function below:

$$\mathcal{L} = \mathcal{L}_2 + \alpha \mathcal{L}_1 \tag{4}$$

where α is a parameter to trade-off the local and global information loss.

4.3 Path Embedding

The motivation for path embedding is as follows, let p be a (acyclic) path containing vertices u and v, we use $E_p(u, v)$ to denote the expectation of traversing from u to v in p with one direction.

Theorem 1. *Let $p = (s_0, s_1, ..., s_n)$ be a path, for any two vertices s_i, s_j in p, $E_p(s_i, s_j) = 1/n$ if $i = 0$ or $i = n$. Otherwise, the following holds:*

$$E_p(s_i, s_j) = \frac{1}{2 \times \max(i, n - i)}. \tag{5}$$

Proof. First consider the case $i = 0$ or $i = n$, s_i is an endpoint of p, only one direction of moving is possible, namely from s_i to s_n or from s_i to s_0. Thus the possibility of moving from s_i to $s_k(s_{n-k}$, resp.) is 1 at step k, and 0 otherwise. Starting from s_i, it takes n steps to traverse all the vertices in p with one direction. The possibility of traversing from s_i to s_j within n steps adds up to 1, so $E_p(s_i, s_j) = 1/n$. Next consider the case $0 < i < n$, s_i is not an endpoint, both direction of moving is possible, each with possibility $1/2$. Analogously, without turning back, s_i will traverse to s_j only once. Starting from s_i, it takes $\max(i, n - i)$ steps to traverse all the vertices in p with one direction. The possibility of traversing from s_i to s_j within $\max(i, n - i)$ steps adds up to $1/2$, so $E_p(u, v) = 1/(2 \times \max(i, n - i))$.

Based on vertices on the path that have the same walking expectation and the prior assumption of consistency [10], we have sufficient reason to embed the same path nodes into the similar Gaussian distributions. Specifically, the path embedding can be seen as depth-first random walks of node2vec [3], specifically, after the Gaussian embedding module is completed, we get the embedding of nodes on coarsening subgraph G_C to get node embeddings of the path set P_C. We directly get μ part of coarsening node embeddings, that is $Y_o(\mu), o \in P_C, o \notin G_C$, via formula $Y_o(\mu) = Y_c(\mu) - \gamma$, where $Y_c(\mu), c \in G_C$ are obtained by Gaussian embedding module, γ is a small random constant. For σ part of coarsening node embeddings, we employ distance function g on each two connected node pairs, one from the Gaussian embedding module and the other from the path set P_C. During the training, only the σ part of coarsening nodes are trainable parameters.

5   Experiments

In this section, we introduce the datasets and baseline algorithms, and then we evaluate our approach on various experiments.

5.1   Datasets

We implement our experiments on four citation network datasets to show the performance of GLP2Gauss. *Cora* dataset consists of 2708 scientific publications and 5429 links across seven classes. *CiteSeer* is a citation graph based on publications in computer science, which consists of 4230 nodes, 5358 edges, and the labels are classified into six categories. *Pubmed* is another citation network, which contains 18230 nodes, 79612 links, three classes. *DBLP* [5] is a bibliography dataset, which contains 17716 nodes,105734 links, 4 classes. All the networks are undirected, and the statistics of four datasets are shown in Table 1.

Table 1. Statistics of four datasets.

	Cora	CiteSeer	Pubmed	DBLP
|Vertices|	2,708	4,230	18,230	17,716
|Edges|	5,429	5,358	79,612	105,734
|Labels|	7	6	3	4

Table 2. AUC results of link prediction.

	Cora	Citeseer	Pubmed	DBLP
GLP2Gauss	**0.948**	**0.952**	**0.969**	0.954
DVNE	0.947	–	–	–
DeepWalk	0.923	0.854	0.913	0.895
Line	0.931	0.928	0.873	0.866
SDNE	0.907	0.881	0.905	0.911
G2G_oh	0.942	0.827	0.963	**0.967**

To evaluate the effectiveness of GLP2Gauss, we introduce five algorithms as our baselines:

- DeepWalk [2] uses short random walks to learn representations for vertices in graphs. We take window size as 10 and walk length as 80, and walk per node 10 times.
- LINE [15] is developed for very large-scale information networks. We use the default parameters mentioned in the paper.

- SDNE [17] is a deep auto-encoder model that learns more structural information representations for nodes. We take the default parameter settings as
  implemented in SDNE [17].
- G2G_oh [5] is a particular case of the Graph2Gauss embedding model for a
  plain graph. We also take the default parameter settings as implemented in
  G2G_oh [5].
- DVNE [6] is a Gaussian embedding model for an undirected graph. Because
  DVNE is designed for large datasets and not available, we only use the result
  of DVNE implemented on the Cora dataset.

5.2 Link Prediction

Link prediction measures the ability of embeddings to meet model assumptions.
Following [5], we randomly choose 20% edges and 20% non-edges from the network as a testing set; the rest of the edges are used as a training set. To have
the same experimental setup, we remove some limitations of the sampling process and set all methods to random sampling. After training, we use each node
embedding to calculate AUC scores for testing edges, and the results are shown
in Table 2.

As shown in Table 2, DVNE utilizes the same distance function g as
GLP2Gauss and achieves the second-highest link prediction performance on
Cora. DeepWalk model shows relatively lower link predictive ability, reflecting
missing some edges result in the loss of necessary indirect connection information
between nodes. For LINE, AUC results are a decline in Pubmed and DBLP; the
main reason behind this is that the model of LINE is challenging to learn non-
linear information of networks. As a Deep model, SDNE learns first and second-
order proximity, achieving a relatively high AUC. However, compared with other
deep models, the predictive power of SDNE is not enough. Since it learns node
embedding as a point-vector rather than a distribution, it cannot reflect more
facets of nodes in the networks. We show that the GLP2Gauss model, as high-
order information explicitly used in modeling, offers the best results on three
datasets so that hidden edges could have a relatively small impact on predictive
ability. For Citeseer dataset, GLP2Gauss does not perform well. In GLP2Gauss,
hidden edges from Citeseer have undermined the necessary connections of nodes
since Citeseer is a small dataset. Thus latent high-order information in network
is hard to be captured by G2G_oh.

5.3 Classification

Our evaluation follows the classification method in [2]: given a dataset, we
increase the rate of training vertices from 1% to 10% to learn a linear SVM
classifier [16] which is employed to predict the rest vertices. We repeat the experiment of a dataset five times and an average score, and we show the results in
Fig. 4.

As shown in Fig. 4, GLP2Gauss has the best performance in four citation
networks when the percentage of training data is relatively low since GLP2Gauss

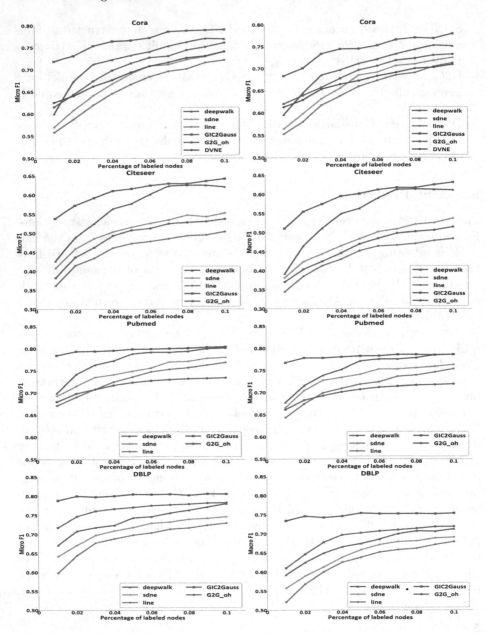

Fig. 4. Micro-F1 and Macro-F1 on four datasets.

can still achieve a high classification score. Both DVNE and G2G_oh keep a relatively steady growth trend, compared with other models. Therefore, we can see that Gaussian embedding methods are more smooth when using different proportions of training sets.

The Line delivers the slightly worse results in most experiments, and the main reason is that non-linear structure in networks are difficult to be captured by LINE. SDNE adopts an auto-encoder structure for learning non-linear structures in networks. Thus SDNE achieves a better result. DeepWalk samples the nodes near the initial node; sufficient information of each node is obtained in this process. Meanwhile, the relationship between nodes and their non-connected nodes is established. Thus DeepWalk achieves excellent classification performance. These results further illustrate the importance of local path (or indirect connectivity). G2G_oh also achieve good results on DBLP and Cora networks, and weak on Citeseer network, because Citeseer has fewer high degree nodes, resulting in less 1-hop information available for each node. Since G2G_oh only consider 1-hop information in their loss function, it is challenging to characterize indirect connectivity between nodes and their non-connected nodes. Moreover, indirect connectivity is implicit in networks so that G2G_oh cannot automatically recognize this information between nodes and their non-connected nodes. Thus the capability of G2G_oh is limited.

Our model performing node embeddings is similar to G2G_oh. However, we still offer the best classification results since we explicitly declare the indirect connectivity in networks and utilize this information. Notably, local path information in networks is essential, and explicit use of this information in the model makes node embeddings retain information. To sum up, GLP2Gauss outperforms DeepWalk, LINE, SDNE, G2G_oh in terms of Micro and Macro F1 on all four datasets. More precisely, HybridGauss is higher than DeepWalk (1.00%–1.28%), LINE (1.04%–1.46%), SDNE (1.02%–1.42%), G2G_oh (1.04%–1.38%). The Micro-F1 and Macro-F1 values of GLP2Gauss are stable at a high level when increasing the percentage of labeled vertices. In these cases, HybridGaussian is not sensitive to the restrictions of the scale of labeled vertices in training. Hence, GLP2Gauss is robust in the classification task.

5.4   Embedding Expectation and Variance

Following [6], we evaluate the uncertainty of node embeddings and find nodes positions are also showing a noteworthy trend. We first select the average value of 10 dimensions with the largest variance as the variance value of each node. Secondly, according to the $\log_{10}$ values of degree of nodes, we divide nodes into ten parts where each part variance is set as the average variance of each part nodes. The expectations of the nodes are divided in the same way, and the Citeseer dataset is removed from this task since the largest degree of nodes in Citeseer dataset is 17. We report the result in Fig. 5.

We can see that nodes with low degree have more uncertainty than others on three datasets. As the degree of nodes increases, the uncertainty of node embeddings decreases gradually. Moreover, the last point at the back of the curve of DBLP has a slightly higher variance than other nodes due to the front of the curve consisting of most of the nodes. As a result, the average variance of these nodes is more stable. On the contrary, the number of nodes behind the curve is usually less than 5 in our dataset. Thus the value of the last point in DBLP is not

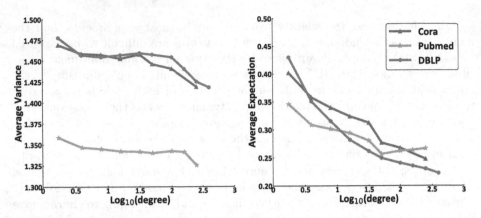

Fig. 5. Expectation and variance of embeddings.

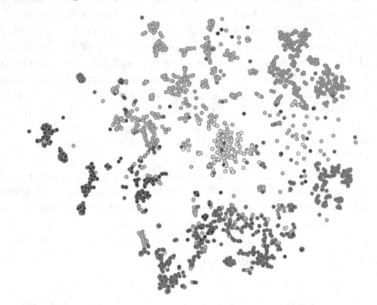

Fig. 6. Visualization of cora in GLP2Gauss (Color figure online)

averaged so that it could have a slightly higher variance. Another point worth noting is that the position of node embeddings has the same tendency as the uncertainty of node embeddings. It reflects that the dimension characteristics of nodes with similar degrees tend to be consistent. In summary, more aspects of the nodes in networks can be reflected by embed nodes into Gaussian distributions.

5.5   Visualization

In this section, we present the visualization of embedding in GLP2Gauss and DeepWalk [2] model by mapping all learned node embeddings in a 2D projected space shown in Fig. 6 where all vertices in color have the same label. It is shown that vertices in the same color are close together. In this sense, GLP2Gauss embedding can characterize the similarity of vertices.

6 Conclusion

In this paper, we present a Gaussian embedding model named GLP2Gauss by introducing our proposed graph structure of a network. We show that GLP2Gauss can better characterize local semantic connectivity than current Gaussian embedding models. In GLP2Gauss, we develop coarsening approaches to sample positive vertices in a network via our proposed structure. We show that GLP2Gauss can characterize the local semantic connectivity of networks instead of a random walk to capture indirect connectivity. We plan to apply GLP2Gauss to more tasks and heterogeneous networks in future work.

Acknowlegements. This work is supported by the National Key Research and Development Program of China (2017YFC0908401) and the National Natural Science Foundation of China (61972455).

References

1. Mikolov, T., Chen, K., Corrado, G., Dean, G.: Efficient estimation of word representations in vector space. In: Proceedings of ICLR (2013)
2. Perozzi, B., Al-Rfou, R., Skiena, S.: DeepWalk: online learning of social representations. In: Proceedings of KDD, pp. 701–710 (2014)
3. Grover, A., Leskovec, J.: node2vec: scalable feature learning for networks. In: Proceedings of KDD, pp. 855–864 (2016)
4. Pan, S., Zhu, X., Zhang, C., Wang, Y.: Tri-party deep network representation. In: Proceedings of IJCAI, pp. 1895–1901 (2016)
5. Bojchevski, A., Günnemann, S.: Deep Gaussian embedding of attributed graphs: Unsupervised inductive, CoRR, abs/1707.03815 (2017)
6. Zhu, D., Cui, P., Wang, D., Zhu, W.: Deep variational network embedding in Wasserstein space. In: Proceedings of KDD, pp. 2827–2836 (2018)
7. Hamilton, W., Ying, R., Leskovec, J.: Representation learning on graphs: Methods and applications, CoRR, abs/1709.05584 (2017)
8. Yang, C., Liu, Z., Zhao, D., Sun, M., Chang, E.: Network representation learning with rich text information. In: Proceedings of IJCAI, pp. 2111–2117 (2015)
9. LeCun, Y., Chopra, S., Hadsell, R., Ranzato, M., Huang, F.: A tutorial on energy-based learning. Predict. Struct. Data, **1** (2006)
10. Zhou, D., Bousquet, O., Lal, T.N., Weston, J., Schölkopf, B.: Learning with local and global consistency. In: Proceedings of NIPS, pp. 321–328 (2003)
11. Cai, H., Zheng, V.M., Chang, K.C.C.: A comprehensive survey of graph embedding: problems, techniques, and applications. IEEE Trans. Knowl. Data Eng. **30**(9), 1616–1637 (2018)
12. Vilnis, L., McCallum, A.: Word representations via Gaussian embedding. In: Proceedings of ICLR (2014)

13. He, S., Liu, K., Ji, G., Zhao, J.: Learning to represent knowledge graphs with Gaussian embedding. In: Proceedings of CIKM, pp. 623–632 (2015)
14. Givens, C.R., Shortt, M.: A class of Wasserstein metrics for probability distributions. Michigan Math. J. **31**(2), 231–240 (1984)
15. Tang, J., Qu, M., Wang, M., Zhang, M., Yan, J., Mei, Q.: LINE: large-scale information network embedding. In: Proceedings of WWW, pp. 1067–1077 (2015)
16. Gunn, S.R.: Support vector machines for classification, and regression. ISIS Tech. Rep. **14**(1), 5–16 (1998)
17. Wang, D., Cui, P., Zhu, W.: Structural deep network embedding. In: Proceedings of KDD, pp. 1225–1234 (2016)
18. Dai, H., et al.: Adversarial attack on graph structured data. In: Proceedings of ICML, pp. 1123–1132 (2018)
19. Dong, Y., Chawla, N., Swami, A.: metapath2vec: scalable representation learning for heterogeneous networks. In: Proceedings of KDD, pp. 135–144 (2018)
20. Ying, R., He, R., Chen, K., Eksombatchai, P., Hamilton, W.L., Leskovec, J.: Graph convolutional neural networks for web-scale recommender systems. In: Proceedings of KDD, pp. 974–983 (2018)
21. Veliakovia, P., Fedus, W., Hamilton, W.L., Lio, P., Bengio, Y., Hjelm, R.D.: Deep graph infomax. CoRR, abs/1809.10341 (2018)
22. Kipf, T.N., Welling, M.: Semi-supervised classification with graph convolutional networks. CoRR, abs/1609.02907 (2016)
23. Velickovic, P., Cucurull, G., Casanova, A., Romero, A., Lio, P., Bengio, Y.: Graph attention networks. CoRR, abs/1710.10903 (2017)
24. Sohn, K.: Improved deep metric learning with multi-class n-pair loss objective. In: Proceedings of NIPS, pp. 1857–1865 (2016)

A Novel Shilling Attack Detection Method Based on T-Distribution over the Dynamic Time Intervals

Wanqiao Yuan[1], Yingyuan Xiao[1(✉)], Xu Jiao[1], Chenchen Sun[1], Wenguang Zheng[1], and Hongya Wang[2]

[1] Tianjin Key Laboratory of Intelligence Computing and Novel Software Technology, Tianjin University of Technology, Tianjin 300384, China
yyxiao@tjut.edu.cn
[2] College of Computer Science and Technology, Donghua University, Shanghai 201620, China

Abstract. The recommendation systems have become an important tool to solve the problem of information overload. However, the recommendation system is greatly fragile as it relies heavily on behavior data of users. It is very easy for a host of malicious merchants to inject shilling attacks in order to control the recommendation results. Some papers on shilling attack have proposed the detection methods, but they ignored experimental performance of injecting a small number of attacks and time overhead. To solve above issues, we propose a novel detection method of shilling attack based on T-distribution over dynamic time intervals. Firstly, we proposed Dynamic Time Intervals to divide the rating history of items into multiple time windows; secondly, the T-distribution is employed to calculate the similarity between windows, and the feature of T-distribution is obvious to detect small samples; thirdly, abnormal windows are identified by analyzing the T value, time difference and rating actions quantity of each window; fourthly, abnormal rating actions are detected by analyzing rating mean of abnormal windows. Extensive experiments are conducted. Comparing with similar shilling detection approaches, the experimental results demonstrate the effectiveness of the proposed method.

Keywords: Shilling attack · Dynamic Time Intervals · T-distribution · Recommendation system

1 Introduction

The recommendation systems [1, 2] have been extensively applied to various e-commerce websites to help users get rid of the trouble of information overload, but it heavily depends on user behavior data (e.g. ratings or clicks of users). Some malicious merchants [3] pour into a large amount of biased user rating profiles in the recommendation systems with the purpose of changing the recommending results for their own profits. These artificially biased user rating profiles are called attack profiles [4–7].

© Springer Nature Switzerland AG 2020
Y. Nah et al. (Eds.): DASFAA 2020 Workshops, LNCS 12115, pp. 225–240, 2020.
https://doi.org/10.1007/978-3-030-59413-8_19

The growing number of shilling attackers has been disturbing the recommendation results, which seriously affects the stability and accuracy of the recommendation systems. To solve the problem, Zhang et al. [8] construct a time series of rating to detect abnormal attack events according to the sample average and sample entropy in each window. But their fixed windows easily overflow attack events. Gao et al. [9, 10] propose two methods for detecting shilling attack. The former is based on fixed time intervals, and the latter adopts a dynamic partitioning method for time series. However, they neglect experimental performance of injecting a small amount of attack profiles.

To address the issue, this paper proposes a novel detection method of shilling attack based on T-distribution on Dynamic Time Intervals. Firstly, we analyze the difference between abnormal users and normal users based on shilling attack features [8–10] and present Dynamic Time Intervals algorithm; in addition, we use T-distribution to calculate the similarity between windows, which has significantly recognition capability of small sample; furthermore, we identify abnormal windows by analyzing the T value, time difference and rating actions quantity of each window; finally, abnormal rating actions are detected by comparing with rating mean of abnormal windows.

The main contributions of this paper are summarized as follows.

- We find that, analyzing statistical features, the incidence rate of injection attack events periods is tiny time intervals throughout entire lifecycle of the item.
- We proposed dynamic time intervals based on the existing shilling attack features to divide the rating history of items into multiple time windows.
- The T-distribution, whose feature is obvious to detect small samples, is employed to calculate the similarity between windows and we analyze the T value, time difference and rating actions quantity of each window to identify abnormal windows.
- Extensive experimental results demonstrate detection method of shilling attack based on T-distribution on dynamic time intervals is obvious to strengthen recommendation system.

The rest of this paper are organized as follows. Section 2 reviews the related work. Section 3 introduce T-distribution and define some notations and concepts used in this paper. Section 4 presents the proposed method. Section 5 evaluates our method through extensive experiments. Section 6 summarize this paper.

2   Related Work

2.1   Attack Profile and Attack Model

The shilling attack [3] main includes push and nuke attacks. The push attacks can enable the recommendation systems to make it easier to recommend target items, and the nuke attacks are to make it more difficult to recommend target items. To avoid being detected, many attack models have been introduced to disguise themselves. Several common attack models are proposed here. Attack profiles contain filler items, selected items, unrated items and target items [4]. Selected items are based on particular needs of the fake users, filler items are randomly selected items to disguise normal users, unrated items are those items with no ratings in the profiles, and target items are the items that attackers attempt to promote or demote. The selected items and filler items strengthened the power of the shilling attack.

These basic elements in attack profiles make up different types of attack models. The three most common attack models are random attack, average attack, and popular attack. The random attack is to randomly select some items from the system for random ratings. Its advantage is that requires little cost but gains great benefits. The average attack is that items of random attack are assigned the corresponding average ratings. Popular attack is an extension of random attack, on this basis, the most popular item in the field is selected to have the highest rating. The model structure is shown in Table 1. In addition, in order to avoid detection, the attacker also adds obfuscated attack, which is more difficult to detect [5–7].

Table 1. Three classical attack models.

Attack type	Push attack	Nuke attack
Random	$I_S = \Phi$, $I_F = r_{ran}$, $I_t = r_{max}$	$I_S = \Phi$, $I_F = I_{ran}$, $I_t = r_{min}$
Average	$I_S = \Phi$, $I_F = r_{avg}$, $I_t = r_{max}$	$I_S = \Phi$, $I_F = I_{ran}$, $I_t = r_{min}$
Bandwagon	$I_S = r_{max}$, $I_F = r_{ran}$, $I_t = r_{max}$	$I_S = r_{max}$, $I_F = r_{ran}$, $I_t = r_{min}$

2.2 Shilling Attack Detection Methods

For robustness of the recommendation systems, many experts have researched some methods to detect shilling attacks. Chirita et al. [11] the earliest proposed the attribute RDMA (Rating Deviation from Mean Agreement), which characterizes the difference of user rating vector, and identified user profiles by average similarity and RDMA. Burke et al. [12] also defined some detection features about user profiles to classify normal and abnormal user. However, their classifications are not obvious. Later, more shilling attack detection methods have been proposed [13]. These methods are mainly divided into four categories [14], including supervised learning model, unsupervised learning model, semi-supervised learning model and statistical analysis method.

- **Supervised learning model:** Through training labeled data, classifier parameters are adjusted to identify attack profiles. Li et al. [15] proposed detection attacks method based on user selecting patterns. Firstly, the method extracted features through user selecting patterns; and then, the method constructed classification based on these features to attack profiles. Fan et al. [16] proposed the method through constructing Bayesian model and analyzing user potential features. However, these methods can only apply to existing specific attack models and need the suitable dataset.
- **Unsupervised learning model:** These models do not require training set. Attack profiles are identified by clustering based on user attribute. Yang et al. [17] proposed unsupervised method of detection attack profiles. The method includes three phases. Firstly, user undirected graph was constructed based on original data and the method

calculated similarity between users by graph mining method to reduce graph; Secondly, partly normal users were excluded by analyzing difference between users; Thirdly, target items were analyzed to detect attack profiles. Zhang et al. [18] proposed methods combination of hidden Markov and hierarchical clustering. Firstly, the method established user rating history model based on hidden Markov and analyzed user preferences to calculate suspicious degree; secondly, hierarchical clustering was employed to gain the attacker set. However, their experiments need strong preliminary knowledge and certain assumptions.

- **Semi-supervised learning model:** The learning methods combine supervised and unsupervised learning. They use a large number of unlabeled and labeled data for pattern recognition. Wu et al. [19] proposed hybrid attack detection method. Firstly, MC-Relief was employed to select shilling attack attribute; Secondly, Semi-supervised Naive Bayes model was used to identify abnormal users. Zhang et al. [20] proposed PSGD to identify abnormal user groups. PSGD uses labeled abnormal groups and unlabeled groups to detect attackers. However, their expensive calculational overhead and complexity are troublesome.

- **Statistical analysis method:** These methods analyze and survey obtained data and information and establish mathematical models to identify abnormal events. These kinds of methods have universality because of the view of items. Zhang et al. [8] proposed a time series detection method based on the sample mean and sample entropy according to the rapidity and purpose of the attacker. However, the time window of this method is fixed, and attack events easily overflow the window. Gao et al. [9, 10] proposed dynamic dividing for time series based on significant points, and then used χ^2 to identify abnormal time window. However, χ^2 heavily depended on the population variance, and experiment performance is not obvious.

3 Preliminaries

T-distribution was first proposed by William Sealy Gosset [21]. T-distribution is dramatically significant for statistics researches to tackle several practical problems through Interval Estimation and Hypothesis Testing, which has significant applications in the product life, fishery, agriculture domain. Compared with the standard normal distribution curve, T-distribution has a lower middle of the curve, and a higher tail of the curve. This feature of T-distribution is significant for us to detect samples injected with a small number of attacks.

In order to facilitate the description below, we have the following notations and definitions:

- I: the set of the entire items.
- U: the set *of* the entire users.
- H': the set of rating actions. Specifically, each rating action $h \in H'$ is represented as $h = \langle h.i, h.u, h.r, h.t \rangle$, where $h.i \in I$ refers to an item, $h.u \in U$ denotes the user that gives $h.i$ a rating, $h.r$ is the rating, and $h.t$ is the time of rating.

Definition 1 (Rating History). For each item $i_k \in I$, a rating history H_k of item i_k is a sequence of rating records formatted as $H_k = h_1 \xrightarrow{h_2.t-h_1.t} h_2 \xrightarrow{h_3.t-h_2.t} \ldots \xrightarrow{h_n.t-h_{n-1}.t} h_n$, where $h_j \in H'$, $0 < j < n$, $h_{j+1}.t > h_j.t$, and $\nexists h_{j'} \in H'$, $s.t. h_j.t < h_{j'}.t < h_{j+1}.t$.

Definition 2 (Item-ratings Time Gaps Series). For each rating history H_k, an item ratings time gaps series is $IRTGS_k = \{(midT_1, gap_1), \ldots, (midT_{n-1}, gap_{n-1})\}$, which corresponds to a time middle series $midT = \{midT_1, midT_2, \ldots, midT_{n-1}\}$, and a time gap series $gap = \{gap_1, gap_2, \ldots, gap_{n-1}\}$. $midT_j = (h_{j+1}.t - h_j.t)/2$, refers to the median of the adjacent timestamps between ratings, and $gap_j = h_{j+1}.t - h_j.t$, refers to the adjacent timestamps gap between ratings.

Definition 3 (Time Window). Suppose H_k is divided into m time intervals, time window series of H_k is $W_k = \{w_1, w_2, \ldots, w_m\}$, each time window $w_x \in W_k$ corresponds to all $h \in H_k$ of the x th time interval. Besides, $w_1 \cup w_2 \cup \ldots \cup w_m = H_k$, $w_1 \cap w_2 \cap \ldots \cap w_m = \emptyset$.

4 The Proposed Method

In order to gain more benefits, attackers inject a number of high (push attacks) and low ratings (nuke attacks) into the target item, in order to the recommendation system easier or harder to recommend the item [8]. Therefore, a common characteristic is that a number of abnormal rating actions will be injected into the target item in a short time interval [10]. That is to say, high rating quantity or low rating quantity will be significantly increased in the attack time interval. According to the shilling attack characteristics mentioned above, we propose TDTI (T-distribution to detect abnormal rating actions on the Dynamic Time Intervals) method.

Our method is mainly based on two features of the attack as preconditions:

1) Attackers inject a number of abnormal rating actions in a short time in order to save costs, and attacks must be very dense [8–10].
2) The incidence rate of injection periods must be tiny time intervals throughout entire lifecycle of the item.

The general idea of the proposed method is as follows: Firstly, we divide the rating history of item into time windows series by DTI (Dynamic Time Intervals); Secondly, the T-distribution is employed to calculate the similarity between windows. Thirdly, Abnormal windows are identified by analyzing the T value, time difference and rating actions quantity of each window; To be more precise, we calculate the ratings mean of abnormal windows, exclude the rating actions less than the mean (push) or the rating actions more than the mean (nuke), and the rest are the abnormal rating actions in attack profiles.

In this section, we will introduce our TDTI method in detail, divided into two modules:

1) We design the DTI partitioning rating history of item into time window series.
2) T-distribution identifies abnormal rating actions.

4.1 Designing the DTI Partitioning Rating History

We aim to partition the rating history of item into time window series and ensure that the abnormal rating actions are divided into one time window in the same period. Based on feature 1 (a number of abnormal rating actions are injected system in a short time), the specific steps are as follows:

1) According to the rating history of an item, we calculate the corresponding IRTGS (Item-ratings Time Gaps Series).
2) Divide IRTGS into two subsequences according to the *gap* maximum in IRTGS.
3) Repeat step2 for the two subsequences again, and record the *midT* values of *gap* maximum in IRTGS as mark points, until the difference between maximum and minimum of *gap* is less than α (α denotes the threshold of difference between *gap* maximum and *gap* minimum in IRTGS and controls the end circulation of DTI). Because of attacks promptness, the α cannot be too large, otherwise the normal rating actions will be integrated into the abnormal window. Meanwhile, the α also cannot be too small, otherwise the abnormal rating actions will be divided into multiple windows. α is related to the rating characteristic and obtained by a large number of experiments.
4) Divide the rating history of the item based on the mark points recorded by step3.

Algorithm 1: DTI(H, α)

Input : H is a Rating history; α is the threshold of difference between *gap* maximum and *gap* minimum in *IRTGS*

Output: Time window series: $W\{w_1, w_2, \ldots, w_m\}, w_j \subset H, 0 < j \leq m$

1: Initialize $gap, midT, W, w$ to four empty lists
2: **for** $i = 0, \ldots, n-1$ **do**
3: | $gap_i \leftarrow h_{i+1}.t - h_i.t, midT_i \leftarrow (h_{i+1}.t + h_i.t)/2$
4: | Add gap_i in gap, Add $midT_i$ in $midT$
5: **end**
6: $S \leftarrow gap, T \leftarrow midT$
7: **PRH**(S, T, α)
8: Add $0, h_n.t + 1$ in $mark$
9: Sort $mark$ in ascend order
10: $l \leftarrow$ obtaining the length of $mark$
11: **for** $i = 0, \ldots, l-1$ **do**
12: | **for** $h_j \in H$ **do**
13: | | **if** $mark_i < h_j.t \leq mark_{i+1}$ **then**
14: | | | Add h_j in w
15: | **end**
16: | Add w in W
17: **end**
18: **return** W

Algorithm 1 describes the process of DTI, which divides the rating history into time window series. Algorithm 2 named PRH (S, T, α) is a sub-algorithm of algorithm 1 and is called by Algorithm 1 at line 7. In the above Algorithm 2, the function GetMax (gap) is responsible for obtaining the maximum element of gap; the function GetMin (gap) returns the minimum element of gap; the function Extract(L, f, e) extracts the elements of subscripts from f to e in the L, where L denotes a list.

In order to illustrate our method more clearly, we randomly selected an item from MovieLens 100k dataset. The IRTGS of the item original data is shown in Fig. 1(a). We injected 50 attack profiles to show clear result, where a host of abnormal rating actions are shown in the red circle of Fig. 1(b). Figure 1(c) shows the mark points are recorded by DTI. We record the horizontal coordinates corresponding to the red vertical line as mark point set in order to divide the rating history. Figure 1(d) shows the cutting of rating history based on the mark points. According to mark points set, the rating history is divided into time window series, and attacks are divided into a time window, which contains tiny normal rating actions.

Algorithm 2: PRH(S, T, α)

Input : S is a gap;

T denotes a $midT$;

α is the threshold of difference between gap maximum and gap minimum in $IRTGS$

Output: the list of mark points: $mark$

1: Initialize $mark$ to an empty list
2: $max \leftarrow$ GetMax(gap), $min \leftarrow$ GetMin(gap)
3: $m \leftarrow$ obtaining the subscript of max
4: $n \leftarrow$ obtaining the length of gap
5: **if** $max - min > \alpha$ **then**
6:    | $sl \leftarrow$ Extract($gap, 0, m$), $sr \leftarrow$ Extract($gap, m + 1, n$)
7:    | $tl \leftarrow$ Extract($midT, 0, m$), $tr \leftarrow$ Extract($midT, m + 1, n$)
8:    | Add $midT[m]$ in $mark$
9:    | $S \leftarrow sl, T \leftarrow tl$
10:   | **PRH(S, T, α)**
11:   | $S \leftarrow sr, T \leftarrow tr$
12:   | **PRH(S, T, α)**
13: **end**
14: **return** $mark$

4.2 Identifying Abnormal Rating Actions Based on T-Distribution

After the abnormal rating actions are detected by the following steps:

1) Calculate the T value between time windows to determine whether each window is similar with the properties of others.
2) Compare the relationship between each T value and its corresponding boundary value, which is shown in Table 2. If the T value beyond the boundary value is

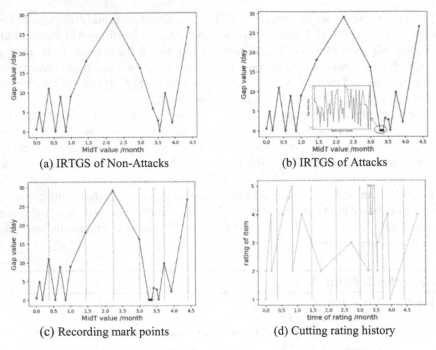

(a) IRTGS of Non-Attacks   (b) IRTGS of Attacks

(c) Recording mark points   (d) Cutting rating history

Fig. 1. DTI partitioning the rating history (Color figure online)

denoted as 1 (dissimilar), and if it is denoted as 0 within the boundary value range (similar). Sequentially, we obtain time window-time window Symmetric Matrix, where its values only have 0 or 1, so in this paper, we call it 0-1 Matrix.

3) Analyze the 0-1 Matrix, and we select the windows whose 1 value quantities are more than the mean of all windows 1 value quantities in rating history to get suspicious window set. Based on the shilling attack features, abnormal window must be minority, so the case that the windows get 1 value is more than the normal windows. Thus, we can easily get suspicious windows by comparing the number of 1 value in each window.

4) Compare each suspicious window obtained in step 3. If time gap of the window is more than the mean time gaps of all windows in the rating history and the rating actions number of the window are less than the mean rating actions quantities of all windows in the rating history, the window is identified as abnormal window.

Table 2. The degree of freedom corresponds to boundary value in the 95% confidence level

Degree of freedom	1	2	3	4	5	6	7	8
Boundary	12.71	4.303	3.182	2.776	2.571	2.447	2.365	2.306

5) Calculate the ratings mean of abnormal windows, exclude the rating actions less than the mean (push) or the rating actions more than the mean (nuke), and the rest are the abnormal rating actions.

The T value is calculated by modified two-sample T-distribution hypothesis testing method [21]. The formula applied to the procedures is as follows:

$$\bar{x}_i^* = \frac{1}{m}\sum_{k=1}^{m} x_{ik}\,(1 \le m \le 5) \tag{1}$$

$$\bar{x}_j^* = \frac{1}{n}\sum_{k=1}^{n} x_{jk}\,(1 \le n \le 5) \tag{2}$$

The formula (1) and (2) are functions of rating action means in a time window. $\bar{x}_i^*$ is the modified mean of the ith time window and $\bar{x}_j^*$ represents modified mean of the jth time window. Since the score range is from 1 to 5 in this context, the max rating difference is 5, leading to that the difference between the rating means of time windows can only be within the range of 5 and the calculation of T value is related to the difference, which is no obvious effect. In order to increase the difference, we improve the function through the m in formula (1) and n in formula (2), refer to the number of rating kinds rather than the number of ratings. x_{i_k} in formula (1) refers to the kth rating in the i window and x_{j_k} in formula (2) is the k th rating in the j window.

$$\bar{x}_i = \frac{m}{g}\bar{x}_i^* \tag{3}$$

$$\bar{x}_j = \frac{n}{h}\bar{x}_j^* \tag{4}$$

The formula (3) is the conversion function of $\bar{x}_i^*$ and $\bar{x}_i$, and formula (4) shows relationship between $\bar{x}_j^*$ and $\bar{x}_j$, where g and h refers to the number of rating in the ith and jth time windows.

$$s_i^2 = \frac{1}{g}\sum_{k=1}^{g}\left(x_{i_k} - \bar{x}_i\right)^2 \tag{5}$$

$$s_j^2 = \frac{1}{h}\sum_{k=1}^{h}\left(x_{j_k} - \bar{x}_j\right)^2 \tag{6}$$

The formula (5) and (6) are variances of a time window.

$$T_{ij} = \frac{\bar{x}_i^* - \bar{x}_j^* - (a_0 - a_i)}{\sqrt{gs_i^2 + hs_j^2}}\sqrt{\frac{mn(m+n-2)}{m+n}} \sim t(m+n-2) \tag{7}$$

T value can be expressed as formula (7), where a_0 refers to the rating mean in entire rating history, and a_i refers to the rating mean of rating history excluding the i th window. $m+n-2$ is the degree of freedom about T value, which corresponds to the boundary value, shown in Table 2. If the degree of freedom is 0, it means that the rating type of both windows is 1, in this case, we default to similar attributes of the two windows.

Subsequently, we take the item selected in the Sect. 4.1 as an example. Table 3 shows the value matrix T_{ij} through T-distribution procedure obtained. And then, compare each T value with the corresponding boundary value to obtain the matrix 0-1 in Table 4. The third window far exceeds the boundary value and third window is suspicious window.

Table 3. T_{ij} value matrix

T_{ij}	1	2	3	4
1	0	2.534	**53.013**	0.157
2	2.683	0	**50.833**	2.43
3	**52.195**	**49.962**	0	**52.132**
4	0.307	2.348	**53.008**	0

Table 4. The matrix 0-1

T_{ij}	1	2	3	4
1	0	0	1	0
2	0	0	1	0
3	1	1	0	1
4	0	0	1	0

5   Experimental Evaluation

5.1   Datasets and Evaluation Metrics

In the experiments, we use public available dataset MovieLens 100 k, which is available on the GroupLens web site and GroupLens Research has collected. The dataset contains 1682 items, 100,000 ratings that were evaluated by 943 users. Each user makes at least 20 ratings, and rating has five ranks from 1 to 5. Statistically analyzing rating time intervals of 1682 items, the mean is 20,029 s and the top 100 is 176.8 s. To facilitate our experiment, we omitted the items with less than 10 rating quantities, because the number of rating was too small to reflect the authenticity of the experiment, and the dataset after filtering is shown in Table 5.

Table 5. Mainly information of the dataset in the experiment

Dataset	Item	User	Rating
MovieLens 100K	1152	943	97,953

We used two indicators [8–10] to evaluate the experimental results: The detection rate in formula (8) is defined as the number of detected attack events (abnormal rating actions) divided by the number of the total attack events. The false alarm rate in formula (9) is defined as the number of normal events (rating actions) that are recognized as attack events divided by the number of all normal events. Here, an event means that a rating action.

$$DetectionRate = \frac{Detected\ Attack\ Events\ Quantity}{Total\ Attack\ Events\ Quantity} \tag{8}$$

$$FalseAlarmRate = \frac{False\ Quantity}{Normal\ Quantity} \tag{9}$$

5.2   Comparative Approaches

At present, there are many methods [11–20] for attacks detection, but we put forward methods based on time and item view, so we compare it with similar methods [8–10] in this paper. The comparison of relevant methods has Gao et al. [9, 10] and Zhang et al. [8] in the simulation experiment on the MovieLens 100k dataset. The following is brief introduction to three methods of them:

- **TS** [8]: The method constructs a time series of rating for an item to compute the features of sample average and sample in each window, and based on duration of attacks, observing the time series of two features can expose attack events, in which sample average is represented by **TS-Ave** and sample entropy is represented by **TS-Ent**.
- **TIC** [9]: The distributions of ratings in diverse time intervals are compared to detect anomaly intervals through the calculation of chi square distribution (χ^2).
- **DP** [10]: According to two features of shilling attacks (the rating of item is always maximum and minimum as well as it takes a very short time to inject large attacks), firstly, this method dynamically divides item-rating time series via significant points, and then, identify abnormal intervals through chi square distribution.

5.3 Experiment Performance

To estimate the performance of our method in different attack quantities, filler sizes, item types and attack models, we use MovieLens dataset to take four item types: fad, fashion, style, and scallop; three attack models: average attack, random attack, and bandwagon attack. Assume that original users are normal users, and attack profiles are those generated from attack models. Based on the features of shilling attacks in Sect. 4, the time gap between attacks in the same period within 1000 s, and the time point of each abnormal rating action is a random point within the attack period. 50 items are randomly selected from 1152 items. The number of attacks is 10, 20, 30, 40, 50 and filler size is set to 1%, 3%, 5%, 7%, 10%, which divided into two categories experience: push attack and nuke attack. We repeat the experiments twenty times, and calculate the detection rate and false alarm according to the mean of the experiments.

Threshold Setting in Experiments. α is the threshold that controls the end circulation of DTI. The impact of α value is related to the item types. Based on the rating features of items in [9], 1,152 items are divided into four types as follows:

- **Fad:** There are 75 items in total. Feature is relatively dense rating time and fewer rating quantities.
- **Fashion**: There are 41 items in total. Feature is relatively dense rating time and larger rating quantities.
- **Style**: There are 250 items in total. Feature is scattered rating time and fewer rating quantities.
- **Scallop**: There are 786 items in total. Feature is continuous topic and larger rating quantities.

We evaluate the impact α on the performance IDTI in four kinds of items. Here, 50 push attacks (filler size is 0) are injected into the system and testing.

Figure 2 shows the effects of α on four item types. We can see that one α value corresponds to a detection rate and a false detection rate, where α value increases from

0, and unit of α is hour (h). In the experiment, we expect the detection rate to be as large as possible and the false alarm rate as small as possible. The most ideal state is that α exists one value, which makes the detection rate reaches 1, and the false alarm rate is 0. We mark ranges of α optimal values in red and structure Table 6. By integrating the intersecting parts of the four item types, we test the best experiment results for four kinds of items when the α value is equal to 0.389 h, as shown in Table 6. In the following experiments, we set α value for DTI as 0.389 h.

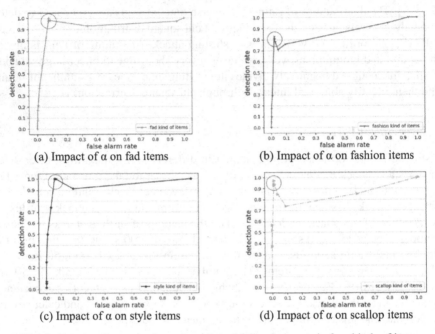

(a) Impact of α on fad items   (b) Impact of α on fashion items

(c) Impact of α on style items   (d) Impact of α on scallop items

Fig. 2. Impact of α on the detection rate and false alarm rate in four kinds of items

Table 6. α optimal value in different kinds of item (α unit: h)

Item types	α	Detection rate	False alarm rate
Fad	0.278–0.389	1	0.0708–0.072
Fashion	0.389–0.444	0.82–0.84	0.023–0.025
Style	0.278–2.78	1	0.062–0.063
Scallop	0.278–0.389	0.96–0.97	0.015
Synthesize	0.389	0.979	0.028

Comparison with Other Approaches. In order to demonstrate the performance improvement of TDTI, we compare TDTI with three methods [8–10] based on the simulation experiments on the MovieLens 100 k dataset. Since the detection methods [8–10] are immune to diverse attack model, the impact of the filler size and attack model isn't considered here. The experiment results are shown in Fig. 3.

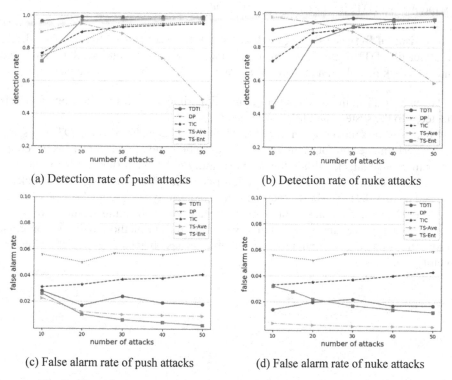

(a) Detection rate of push attacks   (b) Detection rate of nuke attacks

(c) False alarm rate of push attacks   (d) False alarm rate of nuke attacks

Fig. 3. Detection rates and false alarm rate comparison with DP, TIC and TS

- **Detection rate of push attacks:** TDTI has the best experiment performance. We take the Fig. 3(a) as an example, compared with TS-Ent, DP, TIC and TS-Ave, the performance improvements by TDTI are 0.245, 0.215, 0.097 and 0.065 when attack quantity is 10. And when more than 20, TDTI maintains steady state, ranging from 0.99 to 0.992. It is obvious that TDTI has more accurate detection results, because DTI algorithm more accurately divide the abnormal rating actions into a time window, and we employ T-distribution to identify abnormal window. The feature of T-distribution is that small sample has obvious detection effect.
- **Detection rate of nuke attacks:** The experiment performance of TDTI is stable and higher. As shown in Fig. 3(b), when Attack quantity at 10, TDTI is 0.905, and TS-Ave is 0.98, which is 0.075 higher than ours. However, when the attack quantity is 50, TS-Ave is 0.59, and TDTI is 0.97, where we improve 0.38. Because TS-Ave is static window, as the number of attacks increases, it is easy for the attack events to overflow window.

- **False alarm rate:** TDTI is better than TIC and DP. As shown in Fig. 3(c) and Fig. 3(d), TDTI is higher than TS. However, TS has lower detection rate as result of boundedness of static window. TDTI is lower than TIC, DP and relatively stable in the below 0.028. We can see TDTI is more satisfactory for attack detection, because the abnormal windows by TDTI have less normal rating actions.

Comparison of Between Rating Features and Between Attack Models. We detect the effect of TDTI in the four item types: fad, fashion, style and scallop, and three kinds of attack models: random attack, average attack, and bandwagon attack.

We test TDTI in the four kinds of items. Experiment conditions: average attack is injected into the system, and filler size is set to 5%.

Figure 4 shows the detection rate of IDTI on the different item types. IDTI has satisfactory performance on four item types. The detection rate of push attacks is generally higher than nuke attacks. When the attack quantity is 10, the difference between push and nuke is the largest, where the detection rate of fad item injecting push attacks is 0.99 and nuke is 0.91. The best experiment effect of four types is fad, whose push attack gets to 0.99 and nuke attack is up to 0.98. Because the fad rating features are dense rating time and less quantity, which makes it easier for our method to divide abnormal rating actions into a time window, and T-distribution has obvious detection effect on less quantity windows.

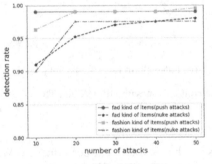

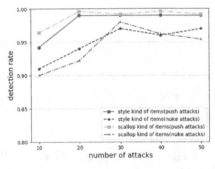

(a) Fad and fashion kinds of items   (b) Style and scallop kinds of items

Fig. 4. Impact of attack size on four kinds of items

We test the impact of the filling size on diverse attack models. Experiment conditions: the attack quantity is set to 10, and the selected item in the Bandwagon attack is 50th because the item always has the topic degree and its rating quantity is largest in the 1152 items.

Figure 5 shows the false alarm rate of TDTI on the diverse attack models. TDTI is unaffected by the attack models and has stable results. The false alarm rate of three models is below 0.028, and the lowest is 0.012. Because TDTI is based on the view of item and consider time factor, our method is immune to different attack models.

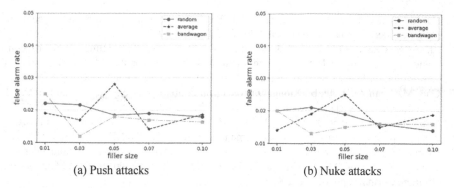

(a) Push attacks   (b) Nuke attacks

Fig. 5. Impact of filler size on attack models

6   Conclusions

Based on the characteristics of the shilling attacks, this paper proposes the DTI algorithm to divide the rating history into multiple time windows, and then the T-distribution is employed to calculate the similarity between windows. Finally, the proposed method identifies shilling attacks by analyzing T value, the time difference and rating actions quantity of each window.

Compared to other methods, the detection rate of our method achieves the desired value, but our false alarm rate still needs to be improved. Therefore, we will focus on how to further decline the false alarm rate in our future work.

Acknowledgment. This work is supported by the National Nature Science Foundation of China (91646117, 61702368) and Natural Science Foundation of Tianjin (17JCYBJC15200, 18JCQNJC00700).

References

1. Yuan, J., Jin, Y., Liu, W., Wang, X.: Attention-based neural tag recommendation. In: Li, G., Yang, J., Gama, J., Natwichai, J., Tong, Y. (eds.) DASFAA 2019. LNCS, vol. 11447, pp. 350–365. Springer, Cham (2019). https://doi.org/10.1007/978-3-030-18579-4_21
2. Xu, K., Cai, Y., Min, H., Chen, J.: Top-N trustee recommendation with binary user trust feedback. In: Liu, C., Zou, L., Li, J. (eds.) DASFAA 2018. LNCS, vol. 10829, pp. 269–279. Springer, Cham (2018). https://doi.org/10.1007/978-3-319-91455-8_23
3. Yi, H., Niu, Z., Zhang, F., Li, X., Wang, Y.: Robust recommendation algorithm based on kernel principal component analysis and fuzzy C-means clustering. Wuhan Univ. J. Nat. Sci. **23**(2), 111–119 (2018). https://doi.org/10.1007/s11859-018-1301-6
4. Kaur, P., Goel, S.: Shilling attack models in recommender system. In: 2016 International Conference on Inventive Computation Technologies. IEEE (2016)
5. Oh, H., Kim, S., Park, S., Zhou, M.: Can you trust online ratings? A mutual reinforcement model for trustworthy online rating systems. IEEE Trans. Syst. Man Cybern.: Syst. **45**(12), 1564–1576 (2015)
6. Cheng, Z., Hurley, N.: Effective diverse and obfuscated attacks on model-based recommender systems. In: RecSys, New York, NY, USA, pp. 141–148 (2009)

7. Yu, J., Gao, M., Rong, W., Li, W., Xiong, Q., Wen, J.: Hybrid attacks on model-based social recommender systems. Physica A **483**(2017), 171–181 (2017)
8. Zhang, S., Chakrabarti, A., Ford, J., Makedon, F.: Attack detection in time series for recommender systems. In: ACM SIGKDD International Conference on Knowledge Discovery and Data Mining, Philadelphia, Pennsylvania, USA, pp. 809–814, August 2006
9. Gao, M., Yuan, Q., Ling, B., Xiong, Q.: Detection of abnormal item based on time intervals for recommender systems. Sci. World J. **2014**, 845–897 (2014)
10. Gao, M., Tian, R., Wen, J., Xiong, Q., Ling, B., Yang, L.: Item anomaly detection based on dynamic partition for time series in recommender systems. PLoS ONE **10**(8), 135–155 (2015)
11. Chirita, P., Nejdl, W., Zamfir, C.: Preventing shilling attacks in online recommender systems. In: Seventh ACM International Workshop on Web Information and Data Management, Bremen, Germany, pp. 67–74 (2005)
12. Burke, R., Mobasher, B., Williams, C., Bhaumik, R.: Classification features for attack detection in collaborative recommender systems. In: ACM SIGKDD International Conference on Knowledge Discovery and Data Mining, Philadelphia, Pennsylvania, USA, pp. 542–547 (2006)
13. Gao, M., Li, X., Rong, W., Wen, J., Xiong, Q.: The performance of location aware shilling attacks in web service recommendation. Int. J. Web Serv. Res. **14**(3), 53–66 (2017)
14. Wang, Y., Qian, L., Li, F., Zhang, L.: A comparative study on Shilling detection methods for trustworthy recommendations. J. Syst. Sci. Syst. Eng. **27**(4), 458–478 (2018). https://doi.org/10.1007/s11518-018-5374-8
15. Li, W., Gao, M., Li, H., Zeng, J., Xiong, Q.: Shilling attack detection in recommender systems via selecting patterns analysis. IEICE Trans. Inf. Syst. **E99.D**(10), 2600–2611 (2016)
16. Fan, Y., Gao, M., Yu, J., Song, Y., Wang, X.: Detection of Shilling attack based on bayesian model and user embedding. In: International Conference on Tools with Artificial Intelligence, pp. 639–646. IEEE (2018)
17. Yang, Z., Cai, Z., Guan, X.: Estimating user behavior toward detecting anomalous ratings in rating systems. Knowl.-Based Syst. **111**(2016), 144–158 (2016)
18. Zhang, F., Zhang, Z., Zhang, P., Wang, S.: UD-HMM: an unsupervised method for shilling attack detection based on hidden Markov model and hierarchical clustering. Knowl.-Based Syst. **148**(2018), 146–166 (2018)
19. Wu, Z., Wu, J., Cao, J., Tao, D.: Hysad: a semi-supervised hybrid shilling attack detector for trustworthy product recommendation. In: Proceedings of the 18th International Conference on Knowledge Discovery and Data Mining, Beijing, China, pp. 985–993 (2012)
20. Zhang, L., Wu, Z., Cao, J.: Detecting spammer groups from product reviews: a partially supervised learning model. IEEE Access **6**(2018), 2559–2567 (2018)
21. Shen, X., Wu, R.: Discussion on t-distribution and its application. Stat. Appl. **4**(4), 319–334 (2015)

Long- and Short-Term Preference Model Based on Graph Embedding for Sequential Recommendation

Yu Liu[1,2], Haiping Zhu[1,2], Yan Chen[1,2(✉)], Feng Tian[1,2], Dailusi Ma[1,2], Jiangwei Zeng[1,2], and Qinghua Zheng[1,2]

[1] School of Computer Science and Technology,
Xi'an Jiaotong University, Xi'an, China
1347740318@qq.com,
{zhuhaiping,chenyan,fengtian,qhzheng}@mail.xjtu.edu.cn,
784352808@qq.com, zxcvbn719270348@163.com
[2] Shaanxi Province Key Laboratory of Satellite and Terrestrial Network Tech. R&D,
Xi'an Jiaotong University, Xi'an, China

Abstract. As sequential recommendation mainly obtains the user preference by analyzing their transactional behavior patterns to recommend the next item, how to mine real preference from user's sequential behavior is crucial in sequential recommendation, and how to find the user long-term and short-term preference accurately is the key to solve this problem. Existing models mainly consider either the user short-term preference or long-term preference, or the relationship between items in one session, ignoring the complex item relationships between different sessions. As a result, they may not adequately reflect the user preference. To this end, in this paper, a Long- and Short-Term Preference Network (LSPN) based on graph embedding for sequential recommendation is proposed. Specifically, item embedding with a complex relationship of items between different sessions is obtained based on graph embedding. Then this paper constructs the network to obtain the user long- and short-term preferences separately, combing them through the fuzzy gate mechanism to provide the user final preference. Furthermore, the results of experiments on two datasets demonstrate the efficiency of our model in Recall@N and MRR@N.

Keywords: Long-termshort-term preference · Sequential recommendation · Graph embedding · Attention networks

This work is supported by National Key Research and Development Program of China (2018YFB1004500), National Natural Science Foundation of China (61877048), Innovative Research Group of the National Natural Science Foundation of China (61721002), Innovation Research Team of Ministry of Education (IRT 17R86), Project of China Knowledge Centre for Engineering Science and Technology, the Natural Science Basic Research Plan in Shaanxi Province of China under Grant No. 2019JM458, MoE-CMCC "Artifical Intelligence" Project No. MCM20190701.

© Springer Nature Switzerland AG 2020
Y. Nah et al. (Eds.): DASFAA 2020 Workshops, LNCS 12115, pp. 241–257, 2020.
https://doi.org/10.1007/978-3-030-59413-8_20

1   Introduction

Recommendation system is widely used in various information service platforms, such as e-commerce, video, news, music, to solve the problem of information overloading caused by massive data and to provide users with personalized recommendation. Most traditional recommendation systems use MF, CF or content-based recommendation [12,21]. It can be concluded as a method that considers the user's historical behavior as a set, and then finds similar people or items to recommend. These methods only consider the user's long-term preference and neglect its dynamic change at different time. Suppose that an electronics enthusiast just bought a computer, then next-item recommendation fits better in related computer accessories (mouse, keyboard, etc.) than watches, bracelets etc. To solve this problem, sequential recommendation was proposed [7], which analyzes the historical behavior sequence of the user with timestamps and dynamically captures user preference to make recommendation results more accurate.

Sequential recommendation is also called session-based recommendation, next-item recommendation. Those methods mainly analyze the user behavior sequence in a certain period of time and obtain the user preference to perform the recommendation [2,31]. And how to mine user's real intention from user behavior sequence is crucial in sequential recommendation. In general, user preference can be divided into user long-term preference (fixed preference) and user short-term preference (current needs). How to combine long- and short-term preference is the main problem of sequential recommendation.

At the beginning, most of the traditional sequential recommendation algorithms used Markov Chain [9] and traditional machine learning methods to model the user's current session [11] to obtain the user short-term preference. Later, more and more deep learning models were applied to sequential recommendation. Recommendation algorithms based on the Recurrent Neural Network (RNN) [2,16] use Gate Recurrent Network (GRU) units to obtain user short-term preference, while CNN-based methods [25] use convolution filter and sliding window to capture it for recommendation. In recent years, there existed some sequential recommendation models that consider both long-term and short-term preference [5,26]. These models use one session as input to train the model, taking into no account that different users have different preferences for items, thus not suitable for personalized recommendation. Though some long-term and short-term combined sequence recommendation models led by RNN [5,18] obtain user preference through GRU or Long Short-Term Memory (LSTM) hidden states, they ignore the different bias of items in the same session, as well as the influence of short-term intent on the model. Meanwhile, most of these models have a problem that the item embedding is constructed in one session or on the entire behavior sequence rather than taking the complex relationship of items between different sessions into consideration. In order to solve this problem, more and more scholars have recently introduced the idea of graph embedding into recommendation systems [27,29], social network [4,22], and knowledge bases [8]. To this end, this paper uses graph embedding to construct item embedding, which is added to the model's original item embedding matrix in a novel way,

initialization, so that the representation of the item has both the complex relationships of static state between sessions and the dynamic implicit relationships in model training. To summarize, the major contributions of this paper are listed as follows:

- Based on graph embedding, the embedding including the complex association relationship between items in different sessions is obtained in addition to the association relationship between the items in one session, and it is added to the item embedding matrix in the model by initialization, making the item embedding constructed more accurate and easier for recommendation.
- The LSPN model is proposed, using GRU and attention to model the user long-term preference, using attention to model the user short-term preference, and finally weighting them through the fuzzy gate mechanism to obtain the user final preference.
- Experiments performed on two datasets show that our model consistently outperforms state-of-the-art methods in terms of Recall@N and MRR@N.

2   Related Work

In this section, we review the related works from three perspectives: sequential recommendation, attention models and graph embedding for deep neural network.

2.1   Sequential Recommendation

Sequential recommendation is mainly based on the implicit feedback (click behavior, purchase behavior, etc.) in the user session to model the user's implicit feedback sequence. There are two types of modeling focus: global models that focused on identifying user's interests in general, and localized models that emphasize user's temporal interests [14]. This type of recommendation is mainly CF. The user-item matrix is decomposed by MF to obtain the user latent factor vector, which means the user long-term preference and item latent factor vector, and then the item recommendation is performed by completing the matrix. In addition, there are traditional machine learning methods such as KNN, recommendation by calculating item similarity, and Markov chain [24]. However, most of these models only consider user long-term preference, and regard it as static, rather than **dynamic**. In recent years, with the rise of neural networks, a variety of deep learning-based models have been applied to sequential recommendation [2,16,25]. Liu Qiao [14] proposed a STAMP model, which, by modeling one session in short-term, considered not only the average preference in one session, but also the combination of long- and short-term preference. Chen Ma et al [15] proposed the HGN model, a sequence recommendation model proposed in 2019. Based on the gated linear unit (GLU) proposed by Dauphin et al., the information of the gating unit used to filter user behavior sequences was modified thus implementing a recommendation. Robin Devooght et al. [5] proposed to perform similar natural language processing operations on the user's item sequence based on GRU. The above models all use one session as input, do not take into account the preference of different users for different item sequences, and make no personalized recommendation.

2.2 Attention Models

With the rise of attention in recent years, considering that different users have different degrees of emphasis on items in a sequence, attention-based sequence recommendation models have also been proposed [32,33]. Meirui Wang et al. [17] proposed a CSRM model, which considers the impact of current sessions and neighborhood sessions on the next-click. TingBai [1] proposed the LSDM model that the user's next click is determined by the user's interest in the product and their needs at different times. Haochao Ying et al. [30] proposed a two-level attention model. However, it lacks the consideration of the last session of the user. And the user long-term preference is to some extent forgettable over time. Therefore, the user long-term preference constructed by this model is not that accurate. In the experimental part, it will be specifically compared that the performance of the long-term preference model proposed in this paper and the data set of the long-term preference model proposed by SHAN.

2.3 Graph Embedding for Deep Neural Network

Graphs are used as inputs to various neural networks to build complex item relationships, such as complex networks, community partitions, and recommendation systems. Graph embedding is gradually used in recommendation systems [27,29]. The ideas of these models are mainly derived from the word2vec idea in NLP. By constructing item graphs, graph-based correlation algorithms [3,34] are used to obtain item representations. Currently, it is mostly used for rough rowing of items, or directly constructing GNN networks for model training. The SR-GNN model proposed by Wu S et al. [28] is to directly construct the conversation graph as the input of the network, and it better represents the global conversation preference through the gate control graph neural network and the soft-attention mechanism. This model does not consider the preference of different users for the session sequence, and differently the model in this paper uses the graph embedding results as an auxiliary for the LSPN. In the model construction, it also maintains the sequence structure of the item to facilitate the dynamic acquisition of user preference.

The main difference between our work and existing approaches is as follows: Firstly, item embedding is constructed based on graph embedding and added to subsequent models by initialization. Secondly, two parallel network structures are constructed for the user long-term preference and short-term preference. More specifically, for user long-term preference, the GRU operation is adopted first, considering that the user preference for items is different, and then the GRU outputs and the user embedding are set as input for attention network to obtain the user long-term preference. Short-term preference is obtained from another attention network with the last session as input. At last, a fuzzy gate mechanism is used to weight the two embedding to obtain the user final preference.

Table 1. List of notations

$\lvert U \rvert$	The number of users
$\lvert I \rvert$	The number of items
S_i	The length of user action sequence
u	The index of a user
i	The index of a item
S_i^u	The item sequence in one session of user u
$\mathbf{U}, \mathbf{I}$	The user and item embedding vector matrices
$\mathbf{u}$	The embedding vector of user u
$\mathbf{i}, I_j, I_j$	The embedding vector of item
L'	The item sequence embedding vector of user u
d	The dimension of embedding vector

3   Methods

3.1   Problem Formulation

The task of sequential recommendation is to predict the next-item i_{t+1} that user will click. The major symbols are listed in Table 1, in this paper, upper case bold letters denote matrices, lower case bold letters denote columns vectors and non-bold letters denote scalars. The input sample of the dataset used is denoted as $(u, \{S_1, S_2, S_3, ..., S_T\})$, where T is the total number of sessions and $S_t = \{i_1, i_2, i_3, ..., i_n\}$, which represents for a user u, his or her action history of item is $\{i_1, i_2, i_3, ..., i_n\}$, where $u \in U$ is a user index, $i_j \in I$ is an item index. Also, regard $Long_{term} = S_1 \cup S_2 \cup ... \cup S_T$ as input to user long-term preference (general intent), and $Short_{term} = S_T$ as short-term preference (current intent) 's input.

3.2   Graph Embedding

Most sequence recommendation models are to automatically learn in the sequence the item's embedding through the network back-propagation [14,15], and some other scholars use the item2vec idea to construct new item embedding [3,34]. However, these methods only consider the relationship between the items in the sequence and do not consider the complex relationship between the items in different sequences, making the built-in item embedding insufficient. In recent years, graphs have been used as inputs to various neural networks to construct complex item relationships [6,19], such as complex networks, community partitioning, and recommendation systems. This paper builds the embedding of each item based on graph embedding, and adds it to the item embedding in the model through initialization. The following details its implementation.

Algorithm 1. Framework of Generation of Item Embedding

Input: $Graph(V, E)$

    Window size w //set $w = |S_i|$

    Embedding size d //set $d =$ item embedding dimensions

    Walk per node γ // Number of random walks to start at each node

    Walk length t // Length of the random walk started at each node

Output: node embedding, matrix of vertex representations $\phi \in R^{|V| \times d}$.

  1: Initialize ϕ;

  2: **for** $i = 0$ to ϕ **do**

  3:    $V' = Shuffle(V)$;

  4:    **for** v_i in V' **do**

  5:        $nodeSequence = RandomWalk(G, v_i, t)$;//generate the $nodesequences$

  6:        $SkipGram(\phi, nodeSequence, w)$; // generate node embedding

  7:    **end for**

  8: **end for**

Construction of Item Graph. User preference is dynamically changed, and behaviors are considered similar in a certain time [27]. Therefore, one session is regard as the input of the item graph to construct an undirected graph. Item graph can be defined as:

$$G = \{V, E\} \tag{1}$$

where, $V = \{i_p | p = 1, ..., |I|\}$ represents node sets in G, $E = \{e_{pq}\} = \{(v_p, v_q) | 0 < p, q \leq |I|\}$ represents edges sets in G.

Generation of Item Embedding. The item graph is isomorphic so that we generate item embedding based on the DeepWalk algorithm proposed by Perozzi B [19] which is often used for the characterization of homogeneous network node. The pseudo-code of DeepWalk for generating Item embedding is listed in Algorithm 1.

3.3 Long- and Short-Term Preference Network(LSPN)

In sequential recommendation, the user preference changes dynamically over time. How to combine dynamic long-term preference with recent preference is the key to predicting clicks on items for the user at the next moment. Considering that the user long-term preference has a certain time-forgetting and item-oriented nature, we finally consider using GRU and attention to build together. For short-term preference, it can also be called current needs, and only the last session S_T needs to be considered. And a simple layer of attention is used directly to model the user short-term preference. Finally, a neural network gate is used to dynamically combine long-term and short-term preference to obtain the user's final preference.

    This section introduces the LSPN model proposed in our paper. Firstly, the **embedding layer** after adding graph embedding is introduced, followed by the modeling of **long- and short-term preferences**, and finally **prediction**. The framework of the entire model is shown in Fig. 1.

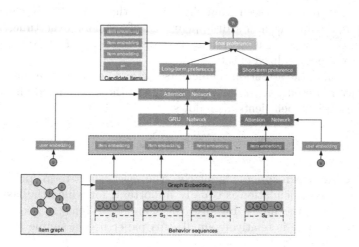

Fig. 1. The architecture of LSPN

- **Embedding Layer.** The input is a user u and his behavior sequence $L = \{S_1, S_2, S_3, ..., S_T\}$. After one-hot, this model uses $\mathbf{U}, \mathbf{I}$ to get user embedding and items embeddings, where $\mathbf{U} \in R^{|U| \times d}$, $\mathbf{I} \in R^{|I| \times d}$. Especially the initialization of $\mathbf{I}$ is the result of the above graph embedding.
- **Long-term Preference.** User long-term preference largely represents user's general preference, but this preference is often dynamic, and in a sense, there is a faded memory in people's perception of the consumption behavior in the past. The traditional recommendation system actually has a fixed representation of the user long-term preference [23]. Although the sequence recommendation in recent years has also dynamically modeled the long-term preference, neither the same items might have different impacts on different users [2], nor the characteristics of human memory loss is taken into consideration, making the obtained user long-term preference inaccurate [30].

Espinaceally, in order to solve these problems, we propose a GRU combined with attention mechanism to simultaneously satisfy different user preference for different items in the sequence and consider the characteristics of the human body's memory fade to dynamically obtain user long-term preference. Espinaceally In order to balance between efficiency and performance, the user item embedding is input into the GRU. The output of each GRU cell and the user's embedding vector u are used as the input of the attention layer. The attention computation is defined as:

$$\alpha_i = \frac{u^T \sigma(W_1 O_i^T + b_1)}{\sum_{i \in L'} u^T \sigma(W_1 O_i^T + b_1)} \qquad (2)$$

where $W_1 \in R^{d \times d}$, $b_1 \in R^{1 \times d}$ are the training parameters of the model, $O_i \in R^{|L'| \times d}$ denotes the output of GRU cell, and $\sigma(\cdot)$ denotes the sigmoid function. α_i represents the attention coefficients of user u within the output of GRU cell. It can be seen from the formula that we not only consider the characteristics of the user's disappearance of memory over time, but also consider that the same items might have different impacts on different users to achieve a personalized effect.

After obtaining the attention coefficients vector α, we compute the long-term preference representation $u_{long-term}$, a sum of the item embedding weighted by the attention coefficients, as follows:

$$u_{long-term} = \sum_{i \in L'} \alpha_i I_i \tag{3}$$

– **Short-term Preference.** Similarly, the user short-term preference in sequential recommendation is equally important, for which the last session, S_T, is taken as the input. Because the number of items in the short-term preference module is relatively small, this paper uses a simple attention to calculate it directly.

The attention computation is defined as:

$$\alpha_j = \frac{u^T \sigma(W_2 I_j^T + b_2)}{\sum_{j \in S_T} u^T \sigma(W_2 I_j^T + b_2)} \tag{4}$$

where $W_2 \in R^{d \times d}$, $b_2 \in R^{1 \times d}$ are the training parameters of the model, $I_j \in R^{|L'| \times d}$ denotes item embedding, and $\sigma(\cdot)$ denotes the sigmoid function. α_j represents the attention coefficients of user u within each item I_j in current session S_T. After obtaining the attention coefficients vector α, we compute the short-term preference representation $u_{short-term}$, a sum of the item embedding weighted by the attention coefficients, as follows:

$$u_{short-term} = \sum_{j \in S_T} \alpha_j I_j \tag{5}$$

– **The LSPN Model.** After we get the user long-term preference and short-term preference separately, the next step is to merge them. Some existing sequence recommendation algorithms also fuse the two, including linear combination [9], Ying H [30] and others who proposed to add long-term preference to the short-term input sequence as a new type of item, dynamically calculating the weight of both through attention, etc. This paper draws on a dynamic weight calculation method **the fuzzy gate mechanism** proposed by Wang M [17] and others in 2019 to obtain the user's final preference. And in the experimental part, SHAN is compared with this model. The specific implementation method is as follows:

$$X_u = \sigma(W_l u_{long-term} + W_s u_{short-term})$$
$$u_{final} = X_u u_{long-term} + (1 - X_u) u_{short-term} \tag{6}$$

– **Prediction.** The user's final preference is obtained through the above three steps. The purpose of sequence recommendation is to predict next item click, so in the end the score of all items need to be calculated. The one with the highest score is the item most likely to be clicked by the user at the next time. The specific implementation method is as follows:

$$score_i = u_{final}I_i \qquad (7)$$

where $I_i \in I$ represents all item are regards as candidate items. The model scores them and gets the recommendation result.

we optimize the proposed model by the Bayesian Personalized Ranking objective [20]: optimizing the pairwise ranking between the positive and negative item as follows:

$$L_{BPR} = -\frac{1}{N_S} \sum_{j=1}^{N_s} log\sigma(r_i - r_j) \qquad (8)$$

where σ is the sigmoid function, r_i and r_j are the ranking scores for sample i and j respectively, N_S is the size of negative samples.

4 Experiments

In this section, we evaluate the proposed model with the state-of-art methods on two real-world datasets.

4.1 Parameter Settings

In order to ensure the fairness of the experiment, the parameters of this experiment are set as follows: the d of the user and item embedding is set as 50, the learning rate without graph embedding introduced is set to 0.001, the learning rate with graph Embedding is set to 0.0005, the optimizer Adam is chosen, and the epoch is selected as 30. For the negative samples of the training set, we randomly sample each user's negative-item for each session, and the number of samples is 99. For the test set, we use all items as the candidate set to score. And the other parameters of the model are adjusted based on the optimal parameters given in the original text to ensure the best results.

4.2 Datasets

In the experiments, we used two real-world datasets, Kaggle's CiaoDVD and Tmall. The Kaggle 's CiaoDVD dataset is about foreign movie rating, and the Tmall contains a large number of e-commerce platform user purchase logs. To ensure the credibility of the experiment, we did some necessary pre-processing on the datasets. Users with a total of no more than 9 interaction records, and products that appear no more than 3 times are filtered out. The original form of the data is $[uid, S_1 : S_2 : ldots : S_n]$. The data set is segmented, in which the first n-1 sessions are train set and the last

Table 2. Detailed informat

	#Users	#Items	Sessions	Sessions per user	Sparsity
CiaoDVD	771	2710	2907	3.770	0.989%
Tmall	17617	48804	54244	3.079	0.041%

is the test set. Segmentation is performed from back to front according to $|S_i|=8$, and the length of the last session is 9, with the last item as the predicted label. The result of dividing the sequence is $[S_1, S_2, ..., S_{n-1}]$ for training, and $[S_n]$ for testing. As the recommendation purpose is to predict the click at the next time, we generate the train set sequences and corresponding labels($[uid, S_1], S_2[0]), ([uid, S_1, S_2], S_3[0]), ..., ([uid, S_1, S_2, ..., S_{n-1}], S_n[0])$). The data format for the test set is: $[uid, S_n[0 : |S_i|]], S_n[|S_i| + 1])$. The statistics of the two datasets after preprocessing are shown in Table 2:

4.3 Evaluation Metrics

This paper uses the following two metrics to evaluate the performance of each method for sequential recommendation problem:

– **Recall@N**: The Recall@N represents the probability that items the user clicked in the test set appear in the top N of the recommended list. Recall@N does not consider the order in which the user actually clicked the item in the recommendation list, as long as it appears in the top N of the recommendation list. The formula is as follows:

$$Recall@N = \frac{n_{hit}}{N} \qquad (9)$$

– **MRR@N**: Another evaluation metric used in the experiment is MRR, which is the average reciprocal of the position number of the item that the user actually clicks on the recommendation list. If the item does not appear in the top N of the recommendation list, MRR is set to 0. MRR@N considers the position of the item in the evaluation recommendation list, and it is a very important metrics in some order-sensitive tasks. The formula is as follows:

$$MRR@N = \frac{1}{N} \sum_{t \in G} \frac{1}{Rank(t)} \qquad (10)$$

4.4 Baselines Methods

To demonstrate the efficiency of our model, we compare it to the following recommendation methods.

– **GRU4Rec** [10]: Hidasi B's RNN-based session recommendation deep learning model proposed in 2016 consists of GRU units, and uses session parallel mini-batch training process, BPR or Top1 loss function during training.

– **STAMP** [14]: A model proposed by Liu Q in 2018. This model uses MLP and attention mechanism to extract long-term and short-term intentions of user in a session for next-item recommendation.
– **NARM** [13]: An RNN-based sequence recommendation model proposed by Li J in 2017. It models the user sequence for GRU first, then adds an attention mechanism to the hidden state of the GRU, and ultimately represents users in different sessions with different characteristics of preference to capture user preference.
– **SHAN** [30]: A model proposed by Ying H in 2018. This model proposes to use a two-layer attention mechanism to obtain user long-term preference and short-term preference respectively, and add the long-term preference to the short-term sequence in the second-level attention.
– **LSPN without GB**: This paper proposes a basic model architecture that does not include graph embedding. The user's overall behavior sequence is modeled separately for long and short periods, and the user's final preference is obtained by weighting the fuzzy gate mechanism.
– **LSPN**: The model is proposed in this paper. Based on the idea of graph embedding, the item embedding layer in the basic model is reconstructed and added to the model for training.

4.5   Experimental Results and Analysis

Table 3 show performances of all methods under the metric of recall and MRR from Top-10, Top-20 in CiaoDVD and Tmall datasets.

Table 3. Performance comparison of LSPN with baselines methods.

	CiaoDVD				Tmall			
	Recall10	MRR10	Recall20	MRR20	Recall10	MRR10	Recall20	MRR20
GRU4Rec	0.0262	0.0092	0.0427	0.0085	0.0147	0.0093	0.0205	0.0094
STAMP	0.0116	0.0039	0.0252	0.0071	0.0082	0.003	0.0159	0.0031
NARM	0.0213	0.0082	0.0262	0.0092	0.0163	0.0066	0.0305	0.0091
SHAN	0.0363	0.0164	0.0395	0.0178	0.0148	0.0049	0.0358	0.0055
LSPN without GB	0.0434	0.0227	0.0667	0.0246	0.0376	0.0079	0.0497	0.0088
LSPN	**0.0757**	**0.0235**	**0.106**	**0.0273**	**0.0405**	**0.0117**	**0.0519**	**0.0125**

As is shown, the proposed model LSPN achieves the best performance on two datasets with all evaluation metrics, which illustrates the superiority of the model. Compared to GRU4Rec, this model only considers the dynamic changes of user preference, and does not consider the process and emphasis of user long-term and short-term preference. The input of the two models of STAMP and NARM are based on a single session, and an attention mechanism is added to consider the different emphasis of the items in the sequence to obtain the main purpose of the user. On the two data sets of CiaoDVD and Tmall, the LSPN is also 8% and 2% higher on Recall@20, respectively. One reason is that

STAMP and NARM did not consider the complex relationship of items in different sessions, making the model's item embedding lack a rich representation. The other reason is that the GRU4Rec, STAMP, and NARM models are built to recommend the next item through the behavior sequence of all users, without considering the preferences of different users for different items, thus neglecting the user's personalized preference. Compared with other models, SHAN is a model that is closer to this paper, but in terms of metrics, the LSPN model is still higher than SHAN. The reason is that SHAN input is consistent with that of this paper. This model lacks the role of the last session for the user long-term preference modeling, and does not consider that the user long-term preference has a certain forgetfulness over time, making the long-term construction of preference not very accurate. In order to verify in detail that the algorithm is different from the experimental results of the algorithm in this paper, we verify on the comparison of the experimental results of the two models separately in the case of only considering long-term preference modeling. Because the CiaoDVD dataset is too sparse and small, the model is easily over-simulated. Therefore, experiments are performed only on the Tmall dataset. The experimental results are shown in Fig. 2. It can be seen that our model LSPN-L is also better than SHAN-L when only long-term preference is considered. Compared with LSPN without GB, the experimental results also prove that graph embedding is indeed useful. The item embedding in the LSPN model is initialized by the graph embedding result. It considers not only the item relationship in the same session, but also the complex relationship between the items in different sessions, making more accurate results.

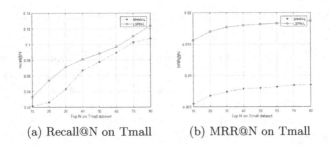

(a) Recall@N on Tmall   (b) MRR@N on Tmall

Fig. 2. Performance comparision between LSPN-L and SHAN-L

5   Model Analysis

In this section, we mainly discuss the impact of **item graphs constructed in different ways, session length** and **embedding size** on the performance of the model.

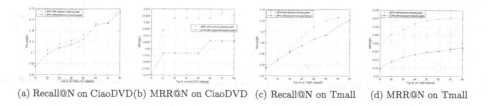

(a) Recall@N on CiaoDVD (b) MRR@N on CiaoDVD (c) Recall@N on Tmall (d) MRR@N on Tmall

Fig. 3. Different methods to construct the item graph on two datasets.

- **Item graphs constructed in different ways**

  This paper adopts two ways of constructing the item graph. One is to borrow from Wang J et al. [27] that the items in one session are more relevant, so one session for each user is used as the basis for constructing the graph. The other is to directly use all the behavior sequences of each user as the basic unit for constructing the graph. In this section, relevant experiments are performed on the two datasets, and the relevant verification and analysis are performed. The result is shown in Fig. 3.

  The experimental results show that the basic unit constructed with one session as the item graph is better than directly using the entire behavior sequence of each user as the basic unit. On both dataset, the results constructed with each user and each session as input are higher in the Recall@20 and MRR@20 metric than constructing the item graph with the entire behavior sequence.

- **Session length**

  During data preprocessing, we use the method of segmenting user behavior sequences. The length of the segmented items is also an important parameter affecting the model. This section mainly explores the effect of session length on experimental results. We show the results under the metric of Recall@20 and MRR@20 on Tmall and CiaoDVD dataset. The results of graph embedding are established by building a separate model, and this section and the next section only consider LSPN with no graph embedding. Then the impact of session length on the model is discussed. The results are show in Table 4.

  As is seen from Table 4, the session segmentation length of different data sets suitable for the model is different. For the CiaoDVD dataset, it is smaller than the Tmall dataset. With session length = 8, recall@20 and MRR@20 are better, and the effect of the model as a whole is relatively large. For Tmall, when the session length = 7, the model achieves a good effect, and the model effect is stable.

- **Embedding size**

  Same as [27], we also study the influence of embedding size in our model. We just show the results under the metric of Recall@20 and MRR@20. And the results are show in Table 5. In the experiment, it is defined that the embedding size of user and item is the same. As shown in Table 5, it can be seen that as the dimension of the embedding size is larger, the CiaoDVD dataset and Tmall dataset are both slowly growing on the Recall@20 and

MRR@20 indicators and are stable until later. It can be understood that the higher the dimension, the more accurate the user and item vectors represented by the model. But when the d is getting bigger, Recall@20 and MRR@20 are getting smaller, and we analyzed that the reason may be that the dimensions are so large that the results are overfitted.

Table 4. Impacts of the session length

Session length($|S_i|$)	CiaoDVD		Tmall	
	Recall20	MRR20	Recall20	MRR20
3	0.0411	0.0035	0.0252	0.0064
4	0.0640	0.0243	0.0295	0.0082
5	0.0675	0.0150	0.0315	0.0091
6	0.0627	0.0192	0.0351	0.0064
7	0.0744	0.0057	0.0413	**0.0116**
8	0.0767	**0.0276**	**0.0498**	0.0088
9	**0.0880**	0.0124	0.0326	0.0049
10	0.0870	0.0066	0.0206	0.0031

Table 5. Impacts of the d

d	CiaoDVD		Tmall	
	Recall20	MRR20	Recall20	MRR20
10	0.0271	0.0157	0.0253	0.0059
20	0.0357	0.0091	0.0377	0.0060
30	0.0476	0.0158	0.0389	0.0073
40	0.0606	0.0086	0.0446	0.0086
50	**0.0667**	0.0246	0.0497	0.0088
60	0.0652	**0.0277**	**0.0539**	**0.0131**
70	0.0600	0.0207	0.0511	0.0097
80	0.0048	0.0194	0.0570	0.0091

6   Conclusion

This paper proposes a Long- and Short-Term Preference Network based on graph embedding for sequential recommendation. The model uses graph embedding to obtain new item embedding, and adds it to the model in an initialized form. Experiments prove that graph embedding is indeed effective. Then, the model proposed in this paper better combines the user long-term and short-term preference, achieving state-of-the-art performance over two datasets. In subsequent experiments, model analysis on more datasets will be performed, together with analysis of the impact of other hyper-parameters on the model.

References

1. Bai, T., Du, P., Zhao, W.X., et al.: A long-short demands-aware model for next-item recommendation. arXiv preprint arXiv:1903.00066 (2019)
2. Hidasi, B., Karatzoglou, A., Baltrunas, L., Tikk, D.: Session-based recommendations with recurrent neural networks. In: Proceedings of ICLR 2015, San Juan, Puerto Rico (2015)
3. Barkan, O., Koenigstein, N.: Item2vec: neural item embedding for collaborative filtering. In: 2016 IEEE 26th International Workshop on Machine Learning for Signal Processing (MLSP), pp. 1–6. IEEE (2016)
4. Cavallari, S., Zheng, V.W., Cai, H., et al.: Learning community embedding with community detection and node embedding on graphs. In: Proceedings of the 2017 ACM on Conference on Information and Knowledge Management, pp. 377–386. ACM (2017)
5. Devooght, R., Bersini, H.: Long and short-term recommendations with recurrent neural networks. In: Proceedings of the 25th Conference on User Modeling, Adaptation and Personalization, pp. 13–21. ACM (2017)
6. Dong, Y., Chawla, N.V., Swami, A.: metapath2vec: Scalable representation learning for heterogeneous networks. In: Proceedings of the 23rd ACM SIGKDD International Conference on Knowledge Discovery and Data Mining, pp. 135–144. ACM (2017)
7. Fang, H., Guo, G., Zhang, D., Shu, Y.: Deep learning-based sequential recommender systems: concepts, algorithms, and evaluations. In: Bakaev, M., Frasincar, F., Ko, I.-Y. (eds.) ICWE 2019. LNCS, vol. 11496, pp. 574–577. Springer, Cham (2019). https://doi.org/10.1007/978-3-030-19274-7_47
8. Guo, S., Wang, Q., Wang, L., et al.: Knowledge graph embedding with iterative guidance from soft rules. In: Thirty-Second AAAI Conference on Artificial Intelligence (2018)
9. He, R., McAuley, J.: Fusing similarity models with markov chains for sparse sequential recommendation. In: ICDM, pp. 191–200 (2016)
10. Hidasi, B., Karatzoglou, A., Baltrunas, L., et al.: Session-based recommendations with recurrent neural networks. arXiv preprint arXiv:1511.06939 (2015)
11. Davidson, J., et al.: The youtube video recommendation system. In: RecSys, pp. 293–296 (2010)
12. Koren, Y., Bell, R., Volinsky, C.: Matrix factorization techniques for recommender systems. Computer **42**(8), 30–37 (2009)
13. Li, J., Ren, P., Chen, Z., et al.: Neural attentive session-based recommendation. In: Proceedings of the 2017 ACM on Conference on Information and Knowledge Management, pp. 1419–1428. ACM (2017)
14. Liu, Q., Zeng, Y., Mokhosi, R., et al.: Stamp: short-term attention/memory priority model for session-based recommendation. In: Proceedings of the 24th ACM SIGKDD International Conference on Knowledge Discovery & Data Mining, pp. 1931–1839. ACM (2018)

15. Ma, C., Kang, P., Liu, X.: Hierarchical gating networks for sequential recommendation. arXiv preprint arXiv:1906.09217 (2019)
16. Quadrana, M., Karatzoglou, A., Hidasi, B., Cremonesi, P.: Personalizing session-based recommendations with hierarchical recurrent neural networks. In: RecSys, pp. 130–137. ACM (2017)
17. Wang, M., Ren, P., Mei, L., Chen, Z., Ma, J., de Rijke, M.: A collaborative session-based recommendation approach with parallel memory modules. In: The 42st International ACM SIGIR Conference on Research and Development in Information Retrieval (SIGIR 2019), pp. 345–354. ACM (2019)
18. Moreira, G.S.P., Jannach, D., da Cunha, A.M.: Contextual hybrid session-based news recommendation with recurrent neural networks. arXiv preprint arXiv:1904.10367 (2019)
19. Perozzi, B., Al-Rfou, R., Skiena, S.: Deepwalk: online learning of social representations. In: Proceedings of the 20th ACM SIGKDD International Conference on Knowledge Discovery and Data Mining, pp. 701–710. ACM (2014)
20. Rendle, S., Freudenthaler, C., Gantner, Z., et al.: BPR: Bayesian personalized ranking from implicit feedback. In: Proceedings of the Twenty-Fifth Conference on Uncertainty in Artificial Intelligence, pp. 452–461. AUAI Press (2009)
21. Sarwar, B.M., Karypis, G., Konstan, J.A., et al.: Item-based collaborative filtering recommendation algorithms. In: WWW, vol. 1, pp. 285–295 (2001)
22. Song, W., Xiao, Z., Wang, Y., et al.: Session-based social recommendation via dynamic graph attention networks. In: Proceedings of the Twelfth ACM International Conference on Web Search and Data Mining, pp. 555–563. ACM (2019)
23. Subramaniyaswamy, V., Logesh, R., Chandrashekhar, M., et al.: A personalised movie recommendation system based on collaborative filtering. Int. J. High Perform. Comput. Netw. 10(1–2), 54–63 (2017)
24. Subramaniyaswamy, V., Logesh, R.: Adaptive KNN based recommender system through mining of user preferences. Wireless Pers. Commun. 97(2), 2229–2247 (2017)
25. Tang, J., Wang, K.: Personalized top-n sequential recommendation via convolutional sequence embedding. In: WSDM, pp. 565–573. ACM (2018)
26. Villatel, K., Smirnova, E., Mary, J., et al.: Recurrent neural networks for long and short-term sequential recommendation. arXiv preprint arXiv:1807.09142 (2018)
27. Wang, J., Huang, P., Zhao, H., et al.: Billion-scale commodity embedding for e-commerce recommendation in alibaba. In: Proceedings of the 24th ACM SIGKDD International Conference on Knowledge Discovery & Data Mining, pp. 839–848. ACM (2018)
28. Wu, S., Tang, Y., Zhu, Y., et al.: Session-based recommendation with graph neural networks. In: Proceedings of the AAAI Conference on Artificial Intelligence, pp. 346–353. AAAI (2019)
29. Xie, M., Yin, H., Wang, H., et al.: Learning graph-based poi embedding for location-based recommendation. In: Proceedings of the 25th ACM International on Conference on Information and Knowledge Management, pp. 15–24. ACM (2016)
30. Ying, H., Zhuang, F., Zhang, F., et al.: Sequential recommender system based on hierarchical attention networks. In: The 27th International Joint Conference on Artificial Intelligence (2018)
31. Zhu, Y., et al.: What to do next: modeling user behaviors by time-LSTM. In: Proceedings of IJCAI 2017, IJCAI, Melbourne, Australia, pp. 3602–3608 (2017)
32. Zhang, S., Tay, Y., Yao, L., et al.: Next item recommendation with self-attentive metric learning. In: Thirty-Third AAAI Conference on Artificial Intelligence (2019)

33. Zhou, C., Bai, J., Song, J., et al.: Atrank: an attention-based user behavior modeling framework for recommendation. In: Thirty-Second AAAI Conference on Artificial Intelligence (2018)
34. Zhou, M., Ding, Z., Tang, J., et al.: Micro behaviors: a new perspective in e-commerce recommender systems. In: Proceedings of the Eleventh ACM International Conference on Web Search and Data Mining, pp. 727–735. ACM (2018)

Event Detection on Literature by Utilizing Word Embedding

Jiyun Chun and Chulyun Kim[✉]

Sookmyung Women's University, Cheongpa-ro 47-gil 100, Yongsan-gu, Seoul 04310, Korea
{wjswldbs2,cykim}@sookmyung.ac.kr

Abstract. The events in literature refer to what the characters are going through, and the story revolves around them. In other words, figuring out events is understanding literature, which is an important concept in assessing the value of it. Event detection is the field of Information Retrieval and previous studies have been mainly based on Automatic Content Extraction (ACE) corpus. However, it costs a lot to make large datasets such as ACE and time consuming because all components like signals for events (entity, event mention, and event argument) are annotated. In addition, it is difficult to apply this large dataset to the special domain of literature. Therefore, we approach event detection in literature as using word embedding. Firstly, we make a set of keywords corresponding to each event, and then we apply Ranking by using word embedding. By utilizing this, this paper will provide a way of finding the sentences relating to the given queries, i.e. the sentences are supposed to be the event. Our method suggests a novel approach in event detection without creating large-scale datasets.

Keywords: Events in literature · Event detection · A set of keywords · Word embedding · Ranking

1 Introduction

Detecting events in a text has a long history in the Information Extraction (IE) field. Most studies were conducted using components based on ACE 2005 [1], such as Entity, Event trigger, and Event Argument [2–4]. Several works proposed event detection as a multi-class classification problem using Convolutional Neural Networks [5–7] or Recurrent Neural Networks [8, 9]. However, these methods require large datasets such as ACE 2005 and neural-based approaches have the problem of error propagation. These kinds of approaches are not a scalable solution for detecting events in various domains because they work only on the basis of events defined in ACE. In addition, building large-scale human-annotated datasets is time-consuming and expensive.

In this paper, we propose a novel method based on ranking for event detection in Literature. Each novel has different expressions of words, such as 'leave' or 'parting'

This research project was supported by Ministry of Culture, Sports and Tourism (MCST) and from Korea Copyright Commission in 2019.

when writers want to express an event 'Break Up'. Thus, if the existing ranking method is applied to event detection, it can find words that exactly match Keywords (i.e., 'Break Up'). Still, it cannot detect similar words (i.e., 'leave', 'parting', etc.) associated with 'Break Up'. In an attempt to solve this problem, we constitute a set of keywords corresponding to each event and then use Ranking with word embedding (we use pre-trained word vectors published by Facebook). When a query corresponding to an event is given, a set of keywords is used to find sentences related to the query. That is, it detects sentences that correspond to the event.

Cbow [10] and Skip-gram [11] word embedding are not available for our event detection method because they cannot handle unknown or Out-Of-Vocabulary (OOV) words and morphologically similar words. As we mentioned, each writer uses different words to describe an event. Therefore, it is impossible to learn all these words. However, for FastText [12], which represents each word as a bag of n-gram, several words are considered to exist within a single word. Learning by considering internal words eliminates the need to learn these words independently; thus FastText word vectors solves the problems mentioned above.

Our approach makes some contributions. We suggest a novel baseline in event detection. Additionally, it is not necessary to make annotated large-scale datasets such as ACE2005.

2   Related Work

2.1   Event Detection with Feature-Based Approaches

Existing methods [2–4] detect events in text based on the following Table 1. To extract events, they use an entity, entity mention, entity trigger, and event argument. Since these components influence the extraction of event, it tends to propagate errors when they are misplaced. In addition, feature-based approaches require rich feature datasets or human-annotated large-scale datasets for achieving high performance. However, creating large datasets is expensive and time-consuming.

Table 1. The definition of components

Component	Definition
Entity	An object or set of objects
Entity mention	A reference to an entity
Event trigger	The word or phrase that expresses an event occurrence
Event argument	Entities that play specific roles in the event

2.2   Event Detection with Neural-Based Approaches

Neural-based approaches have emerged as a way to compensate for the limitations of feature-based methods. Recent studies [5–7] use Convolutional Neural Networks for event detection as a task of multi-class classification. Some studies [8, 9] propose Recurrent Neural Networks. They regarded texts as sequential data, and event triggers and arguments were simultaneously predicted to discover the event. On the contrary, Neural-based methods have the limitation that they ignore semantic relationships in a text.

2.3 Facebook Pre-trained Word Vectors

We chose pre-trained word vectors published by Facebook, trained on Common Crawl, and Wikipedia using FastText. Pre-trained word vectors in Korean are much smaller in size than in English, so we used word vectors in English for better performance. For this reason, the collected novels were translated using Neural Machine Translation (NMT) Papago before the experiment was conducted.

3 Event Definition

To categorize the events that appear in the novel, we first identified the corresponding parts of the events, then we defined and classified them (see Table 2). Emotions such as 'Love', which often appear in Korean literature, are also defined as an event. In addition, there is a different context for the same event. Therefore, we defined them separately. For example, for the context of separation, the word 'parting' is used about death and 'breakups' about relationships. Each event consists of keywords (see Table 3). The criteria for selecting sets of keywords are words that frequently appear in literature for each event.

Table 2. Categorized events

Event	Definition
1. 만남(Meeting)	탄생(Birth), 사랑(Love), 만남(Meeting), 재회(Reunion)
2. 갈등(Conflict)	싸움(Argument), 질투(Jealousy), 오해(Misunderstanding), 살인(Murder)
3. 이별(Separation)	이별(Breakup), 죽음(Death), 가출(Leaving the house), 이사(Moving), 피난(Evacuation), 입양(Adopting), 병(Illness), 자살(Suicide)

4 Event Detection

Each novel is composed of sentences. In other words, a novel is a collection of sentences. We split a sentence into words.

$$Sen_1, \ Sen_2, \ Sen_3, \ldots \ldots, \ Sen_n \qquad (1)$$

Table 3. Sets of keywords

Event	Set of Keywords
탄생(Birth)	탄생(birth), 태어나다(born)
만남(Meeting)	만나다(meet), 만남(meeting)
싸움(Argument)	싸움(quarrel), 싸우다(fight), 폭력(violence)
살인(Murder)	살인(murder), 죽이다(kill)
이별(Parting)	이별(parting), 이별하다(part), 작별(farewell), 떠나다(leave)
죽음(Death)	죽은(dead), 죽다(die), 죽음(death)
자살(Suicide)	죽음(death), 죽다(die), 자살(suicide)
…	…

$$Sen_i = \{word_{i1}, word_{i2}, \ldots, word_{ij}\} \tag{2}$$

Each event $event\_query_\alpha$ consists of keywords $query_{\alpha 1}, query_{\alpha 2}, \ldots, query_{\alpha k}$.

$$event\_query_\alpha = \{query_{\alpha 1}, query_{\alpha 2}, \ldots, query_{\alpha k}\} \tag{3}$$

The cosine similarity between queries and word vectors is the score.

$$\cos \theta = \frac{v_{query_{\alpha_t}} \cdot v_{word_{ij}}}{|v_{query_{\alpha_t}}| \times |v_{word_{ij}}|} \tag{4}$$

$$score_{\alpha_t im} = \text{cosine similarity} (query_{\alpha_t}, word_{im}) \tag{5}$$

X is a set of max_total_score and Y is a set of avg_total_score. $\in_{max\_total\_score}$ is the threshold for X and $\in_{avg\_total\_score}$ is the threshold for Y (β, $\gamma = 0.8$).

$$\in_{max\_total\_score} = \beta \max (X) \tag{6}$$

$$\in_{avg\_total\_score} = \gamma \max (Y) \tag{7}$$

Event Detection Algorithm

Input: Sen_i , $event\_query_\alpha$

1: Split Sen_i into words.
2: Compute Cosine similarity between two-word vectors each query and
   every word in Sen_i. This step repeats k times.
3: If $n(Sen_i) \geq 4$, $total\_score_{\alpha_t i}$ for Sen_i is the sum of the top-4 $score_{\alpha_t im}$.
   (If $n(Sen_i) < 4$, only the largest $score_{\alpha_t im}$ is the $total\_score_{\alpha_t i}$ for Sen_i.)
4: If $n(event\_query_\alpha) \geq 2$, $max\_total\_score_i$ for Sen_i is $max(\{total\_score_{\alpha_1 i},$
   $total\_score_{\alpha_2 i}, ..., total\_score_{\alpha_k i}\})$ and
   $avg\_total\_score_i$ for Sen_i is $avg(\{total\_score_{\alpha_1 i}, total\_score_{\alpha_2 i}, ...,$
   $total\_score_{\alpha_k i}\})$.
5: If $n(Sen_i) \geq 4$, extract Sen_i that both $max\_total\_score_i$ and $avg\_total\_score_i$
   are above the threshold $\epsilon_{max\_total\_score}$ and $\epsilon_{avg\_total\_score}$.
   If $n(Sen_i) < 4$, extract Sen_i that either $max\_total\_score_i$ or $avg\_total\_score_i$
   is above the threshold 1.5.

Output: Sen_i that $event\_query$ (event) occurs

---

5   Experiment

In this paper, we show three events that mainly occur in Korean Literature.

5.1   Event 'Birth' (Test Data: 'Salmosa' by Bumsun Lee)

The experiment was conducted when $event\_query_\alpha$ is 'Birth' and $query$ is 'birth' and
'born'. We draw trajectories for $query$ 'birth' and 'born' (see Fig. 1).

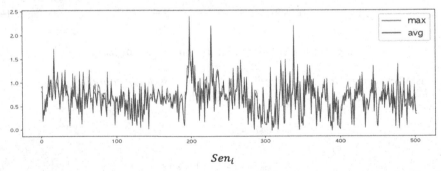

Sen_i

Fig. 1. Trajectories for the event 'Birth'. The red trajectory is $max\_total\_score_i$ for Sen_i and the blue trajectory is $avg\_total\_score_i$ for Sen_i. It shows that 200~230 (i) sections and 320~330 sections are relatively larger than other sections. The result is consistent with sentences in which the event 'Birth' occurs. (Color figure online)

We extract Sen_i by applying threshold (see Fig. 2.)

Before event detection	After event detection
...	...
그의 어머니조차도 그의 아버지를 분명히 모르고 있는 것이라 하였다. 그처럼 살모사의 출생은 그 잉태부터가 기구한 것이었다. (Even his mother said she didn't know his father clearly. The birth of Salmosa was a sinister thing from his birth)	그의 어머니조차도 그의 아버지를 분명히 모르고 있는 것이라 하였다. 그처럼 살모사의 출생은 그 잉태부터가 기구한 것이었다. (Even his mother said she didn't know his father clearly. The birth of Salmosa was a sinister thing from his birth)
...	
그렇게 잉태하여 세상에 태어난 애가 바로 살모사였던 것이다. 궁씨 문중에서는 그날 밤 이야기를 여인에게서 자세히 들은 후 그 애를 호적에 넣었다. (It was Salmosa who was born with such birth. At the palace gate, after hearing the story from ...)	그렇게 잉태하여 세상에 태어난 애가 바로 살모사였던 것이다. 궁씨 문중에서는 그날 밤 이야기를 여인에게서 자세히 들은 후 그 애를 호적에 넣었다. (It was Salmosa who was born with such birth. At the palace gate, after hearing the story from ...)
...	...

Fig. 2. Detecting sentences that the event 'Birth' occurs. Green means sentences corresponding to the event 'Birth', and red means a prediction after the event is detected. Translated text are given in brackets.

5.2 Event 'Meeting' (Test Data: 'Seoul, 1964 Winter' by Seungok Kim)

The experiment was conducted when $event\_query_\alpha$ is 'Meeting' and $query$ is 'meet' and 'met'. We draw trajectories for $query$ 'meet' and 'met' (see Figs. 3, 4).

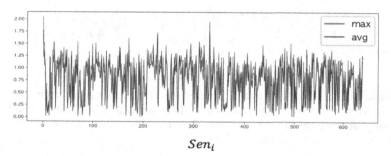

Sen_i

Fig. 3. Trajectories for the event 'Meeting'. The red trajectory is $max\_total\_score_i$ for Sen_i and the blue trajectory is $avg\_total\_score_i$ for Sen_i. It shows that 1~10 (i) sections and 320~330 sections are relatively larger than other sections. The result is consistent with sentences in which the event 'Meeting' occurs. (Color figure online)

5.3 Event 'Death' (Test Data: 'Seoul, 1964 Winter' by Seungok Kim)

The experiment was conducted when $event\_query_\alpha$ is 'Death' and $query$ is 'dead', 'death', and 'die'. We draw trajectories for $query$ 'dead', 'death', and 'die' (see Figs. 5, 6).

Before event detection	After event detection
1964년 겨울을 서울에서 지냈던 사람이라면 누구나 알고 있겠지만, 밤이 되면 거리에 나타나는 선술집 ┄┄ 오뎅과 군참새와 세 가지 종류의 술등을 팔고 있고, 얼어붙은 거리를 휩쓸며 부는 차가운 바람이 펄럭거리게 하는 포장을 들치고 안으로 들어서게 되어 있고, 그 안에 들어서면 카바이드 불의 길쭉한 불꽃이 바람에 흔들리고 있고, 열색한 군용(軍用) 잠바를 입고 있는 중년 사내가 술을 따르고 안주를 구워 주고 있는 그러한 선술집에서, 그날밤, 우리 세 사람은 우연히 만났다. 우리 세 사람이란 나와 도수 높은 안경을 쓴 안(安)이라는 대학원 학생과 정체를 알 수 없었지만 요컨대 가난뱅이라는 것만은 분명하여 그의 정체를 꼭 알고 싶다는 생각은 조금도 나지 않는 서른 대여섯 살짜리 사내를 말한다. (As anyone who spent the winter of 1964 in Seoul would know, the middle-aged man who sold Odeng, a military sparrow and three kinds of liquor lamps that appear on the street at night, lifted up a package that made the cold wind flutter through the icy streets, and inside, the long flames of carbide fire were swaying in the wind, and the middle-aged man wearing a dyed military jacket was drinking wine that day. The three of us refer to me and a graduate student named in high-caliber glasses, but in short, a man aged 35 and 6 who is clearly poor and doesn't have the slightest idea of wanting to know him.) "함께 있어 주십시오." 사내가 말했다. 우리는 승낙했다. "멋있게 한번 써 봅시다."라고 사내는 우리와 만나 후 처음으로 웃으면서, 그러나 여전히 힘없는 음성으로 말했다. ("Please stay with me," said the man. We agreed. "Let's write it nicely," said the man, smiling for the first time since he met us, but still in a weak voice.)	1964년 겨울을 서울에서 지냈던 사람이라면 누구나 알고 있겠지만, 밤이 되면 거리에 나타나는 선술집 ┄┄ 오뎅과 군참새와 세 가지 종류의 술등을 팔고 있고, 얼어붙은 거리를 휩쓸며 부는 차가운 바람이 펄럭거리게 하는 포장을 들치고 안으로 들어서게 되어 있고, 그 안에 들어서면 카바이드 불의 길쭉한 불꽃이 바람에 흔들리고 있고, 열색한 군용(軍用) 잠바를 입고 있는 중년 사내가 술을 따르고 안주를 구워 주고 있는 그러한 선술집에서, 그날밤, 우리 세 사람은 우연히 만났다. 우리 세 사람이란 나와 도수 높은 안경을 쓴 안(安)이라는 대학원 학생과 정체를 알 수 없었지만 요컨대 가난뱅이라는 것만은 분명하여 그의 정체를 꼭 알고 싶다는 생각은 조금도 나지 않는 서른 대여섯 살짜리 사내를 말한다. (As anyone who spent the winter of 1964 in Seoul would know, the middle-aged man who sold Odeng, a military sparrow and three kinds of liquor lamps that appear on the street at night, lifted up a package that made the cold wind flutter through the icy streets, and inside, the long flames of carbide fire were swaying in the wind, and the middle-aged man wearing a dyed military jacket was drinking wine that day. The three of us refer to me and a graduate student named in high-caliber glasses, but in short, a man aged 35 and 6 who is clearly poor and doesn't have the slightest idea of wanting to know him.) … "함께 있어 주십시오." 사내가 말했다. 우리는 승낙했다. "멋있게 한번 써 봅시다."라고 사내는 우리와 만나 후 처음으로 웃으면서, 그러나 여전히 힘없는 음성으로 말했다. ("Please stay with me," said the man. We agreed." Let's write it nicely," said the man, smiling for the first time since he met us, but still in a weak voice.) …

Fig. 4. Detecting sentences that the event 'Meeting' occurs. Green means sentences corresponding to the event 'Meeting', and red means a prediction after the event is detected. (Color figure online)

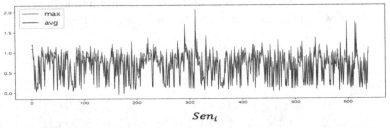

Fig. 5. Trajectories for the event 'Death'. The red trajectory is $max\_total\_score_i$ for Sen_i and the blue trajectory is $avg\_total\_score_i$ for Sen_i. It shows that 280~310 (i) sections and 590~610 sections are relatively larger than other sections. The result is consistent with sentences in which the event 'Death' occurs. (Color figure online)

6 Result

The reason for having low precision, recall, and F-score for the event 'Meeting' (see Table 4.) is that it is mostly translated as 'met', so cosine similarity of 'met' and 'meet'(or 'meeting') is not relatively high. (cosine similarity between them is 0.6934439).

Before event detection	After event detection
...	...
"말씀드리고 싶은 게 있는데요." 마음씨 좋은 아저씨가 말하기 시작했다. "들어 주시면 고맙겠습니다.……오늘 낮에 제 아내가 죽었습니다. 세브란스 병원에 입원하고 있었는데……." ("I'd like to say something." A good-natured uncle began to speak. "I'd appreciate it if you could listen.….my wife died this afternoon. I was in Severance Hospital and……")	"말씀드리고 싶은 게 있는데요." 마음씨 좋은 아저씨가 말하기 시작했다. "들어 주시면 고맙겠습니다.……오늘 낮에 제 아내가 죽었습니다. 세브란스 병원에 입원하고 있었는데……." ("I'd like to say something." A good-natured uncle began to speak. "I'd appreciate it if you could listen.….my wife died this afternoon. I was in Severance Hospital and……")
"급성 뇌막염이라고 의사가 그랬습니다. 아내는 옛날에 급성 맹장염 수술을 받은 적도 있고, 급성 폐렴을 앓은 적도 있다고 했습니다만 모두 괜찮았는데 이번의 급성엔 결국 죽고 말았습니다.……죽고 말았습니다." (The doctor said it was an acute meningitis. My wife said she had surgery for acute appendicitis and had acute pneumonia in the past and everything was fine, but at this acute moment, she ended up dying.……my wife died.")	"급성 뇌막염이라고 의사가 그랬습니다. 아내는 옛날에 급성 맹장염 수술을 받은 적도 있고, 급성 폐렴을 앓은 적도 있다고 했습니다만 모두 괜찮았는데 이번의 급성엔 결국 죽고 말았습니다.……죽고 말았습니다." (The doctor said it was an acute meningitis. My wife said she had surgery for acute appendicitis and had acute pneumonia in the past and everything was fine, but at this acute moment, she ended up dying.……my wife died.")
"아내의 시체를 병원에 팔았습니다. … . 아내가 누워 있을 시체실이 있는 건물을 알아보려고 했습니다만 어딘지 알 수 없었습니다." ("I sold my wife's body to the hospital. I tried to identify the building with the body room where my wife would lie, but I couldn't figure out where it was.")	"아내의 시체를 병원에 팔았습니다. … . 아내가 누워 있을 시체실이 있는 건물을 알아보려고 했습니다만 어딘지 알 수 없었습니다." ("I sold my wife's body to the hospital. I tried to identify the building with the body room where my wife would lie, but I couldn't figure out where it was.")
다음날 아침 일찍 안이 나를 깨웠다. "그 양반 역시 죽어 버렸습니다." 안이 내 귀에 입을 대고 그렇게 속삭였다. (Early the next morning, Ahn woke me up. "He's dead, too." Ahn whispered so with his mouth on my ear.)	다음날 아침 일찍 안이 나를 깨웠다. "그 양반 역시 죽어 버렸습니다." 안이 내 귀에 입을 대고 그렇게 속삭였다. (Early the next morning, Ahn woke me up. "He's dead, too." Ahn whispered so with his mouth on my ear.)
"난 그가 죽으리라는 것을 알고 있었습니다." 안이 말했다. ("I knew he was going to die,"said Ahn.)	"난 그가 죽으리라는 것을 알고 있었습니다." 안이 말했다. ("I knew he was going to die,"said Ahn.)
"난 그 양반이 죽으리라는 짐작도 못 했으니까요." ("I didn't expect him to die.")	"난 그 양반이 죽으리라는 짐작도 못 했으니까요." ("I didn't expect him to die.")
	...

Fig. 6. Detecting sentences in which the event 'Death' occurs. Green means sentences corresponding to the event 'Death', and red means a prediction after the event is detected. (Color figure online)

Table 4. Performance on event detection

Event	Precision	Recall	F-score
Birth	66.7%	100%	80.023%
Meeting	50%	50%	50%
Death	66.7%	60%	63.173%

7 Conclusion

It is meaningful that our paper proposes a new baseline in event detection, unlike the previous studies using feature-based and neural-based methods based on ACE2005. Furthermore, we reduce the annotation burden to make a large-scale dataset for event detection. Additional experiments are currently being conducted, such as varying weights for the query and applying our detection algorithm after extracting sentences that are depicted as actually happening using Deep Learning. We hope this initial approach inspires further research by many researchers in the future.

References

1. Grishman, R., Westbrook, D., Meyers, A.: Nyus English ace 2005 system description. In: ACE 2005 Evaluation Workshop (2005)
2. Li, Q., Ji, H., Huang, L.: Joint event extraction via structured prediction with global features. In: ACL, pp. 73–82 (2013)
3. Li, P., Zhu, Q., Zho, G.: Argument inference from relevant event mentions in Chinese argument extraction. In: ACL (2013)
4. Hong, Y., Zhang, J., Ma, B., Yao, J., Zhou, G., Zhu, Q.: Using cross-entity inference to improve event extraction. In: ACL, pp. 1127–1136 (2011)
5. Zeng, D., Liu, K., Lai, S., Zhou, G., Zhao, J.: Relation classification via convolutional deep neural network. In: COLING, pp. 2335–2344 (2014)
6. Nguyen, T.H., Grishman, R.: Event detection and domain adaptation with convolutional neural networks. In: ACL-IJCNLP (2015)
7. Chen, Y., Xu, L., Liu, K., Zeng, D., Zhao, J.: Event extraction via dynamic multi pooling convolutional neural networks. In: Proceedings of the 53rd Annual Meeting of the Association for Computational Linguistics and the 7th International Joint Conference on Natural Language Processing (ACL-IJCNLP), vol. 1, pp. 167–176 (2015)
8. Nguyen, T.H., Cho, K., Grishman, R.: Joint event extraction via recurrent neural networks. In: NAACL, pp. 300–309 (2015)
9. Feng, X., Huang, L., Tang, D., Qin, B., Ji, H., Liu, T.: A language-independent neural network for event detection. In: Proceedings of the 54th Annual Meeting of the Association for Computational Linguistics, vol. 2, pp. 66–71 (2016)
10. Mikolov, T., Sutskever, I., Chen, K., Corrado, G., Dean, J.: Efficient estimation of word representations in vector space. In: ICLR Work-shop Papers (2013)
11. Mikolov, T., Sutskever, I., Chen, K., Corrado, G., Dean, J.: Distributed representations of words and phrases and their compositionality. In: NIPS, pp. 3111–3119 (2013)
12. Bojanowski, P., Grave, E., Joulin, A., Mikolov, T.: Enriching word vectors with subword information. In: TACL, pp. 135–146 (2017)

User Sequential Behavior Classification for Click-Through Rate Prediction

Jiangwei Zeng[1,2], Yan Chen[1,2], Haiping Zhu[1,2(✉)], Feng Tian[1,2], Kaiyao Miao[1,2], Yu Liu[1,2], and Qinghua Zheng[1,2]

[1] School of Computer Science and Technology, Xi'an Jiaotong University, Xi'an, China
zxcvbn719270348@163.com, miaoky814@xjtu.edu.cn, 1347740318@qq.com, {zhuhaiping,chenyan,fengtian,qhzheng}@mail.xjtu.edu.cn
[2] Shaanxi Province Key Laboratory of Satellite and Terrestrial Network Tech. R&D,Xi'an Jiaotong University, Xi'an, China

Abstract. The user's behavior sequence can well reflect the user's interest and intention, but there is less research on the use of user behavior sequence in the field of CTR prediction. Therefore, introducing the idea of sequence recommendation into CTR prediction is an exciting idea. Aimless browsing by users is a common phenomenon in many recommended scenarios (e-commerce, music, video streaming), which has not been paid attention to in previous research. To this end, this paper introduces the concept of user browsing status, and divides it into Discover and Intent, which respectively represent the user's unintentional status and intentional status. A new framework named User Status Recognition Framework (USRF) is proposed to solve this problem. USRF can perform both CTR prediction and next-item recommendation. The framework captures the weights of two user status from the current user historical behavior sequence, models the two different status separately to mine user interests, and combines the captured user status to make more accurate recommendations. In addition, in order to solve the problem that less attention has been paid to the complex connection between candidate items and interacted items in the previous sequence recommendation research, this paper uses the attention mechanism to model the relationship between candidate items and interacted items, implements a simple model for CTR prediction based on USRF. Experiments on three different scene datasets show good results on both the AUC and F1-score, proving the advantages of the framework.

Keywords: Sequence recommendation · CTR · User status recognition

This work is supported by National Key Research and Development Program of China (2018YFB1004500), National Natural Science Foundation of China (61877048), Innovative Research Group of the National Natural Science Foundation of China (61721002), Innovation Research Team of Ministry of Education (IRT 17R86), Project of China Knowledge Centre for Engineering Science and Technology, the Natural Science Basic Research Plan in Shaanxi Province of China under Grant No. 2019JM458, MoE-CMCC "Artifical Intelligence" Project No. MCM20190701.

Y. Nah et al. (Eds.): DASFAA 2020 Workshops, LNCS 12115, pp. 267–280, 2020.
https://doi.org/10.1007/978-3-030-59413-8_22

1   Introduction

Recommendation systems have been widely used in the real world. For e-commerce sites, video sites and information flow sites, recommendation system is an effective tool to solve the problem of information overload, it makes it easy for users to find what they want. CTR (click-through rate) is used to predict the user's click probability on recommended items or advertisements, so the performance of the CTR prediction model directly affects the company's earnings. With the rapid development of deep learning in the fields of computer vision, speech recognition, and natural language processing, a large number of CTR prediction models based on deep learning have been proposed. More and more companies no longer use the early LR + GBDT model, but use the deep learning model and apply it to the production environment, and have made huge profits.

In the CTR prediction problem, due to the natural sparseness of the user and item features, learning the combined features has become a difficult point. Since the appearance of the FM [1] model, a large number of models have begun to study how to efficiently extract combined features, such as FFM [3], AFM [4], DeepFM [2], etc. But these studies treat the user's behavior as an independent signal, emphasizing the modeling of the complex relationship between the user and the item to be predicted. In fact, the user's behavior is sequential and related. The user's previous interaction history will have a profound impact on the user's behavior at the next moment, so the user's sequential behavior needs to be modeled.

In fact, modeling user's sequence behavior has become a new research direction, that is, sequential recommendation (also called session-based recommendation). Sequential recommendation emphasizes dynamically capturing user's intention and provides next-item recommendation [5] through user's recent behavior sequence. Recently, some studies have begun to focus on how to use the user's behavior sequence for CTR prediction. These studies are mainly carried out by the Alibaba Group [10–12]. However, the field of sequentaial recommendation has more effective and complete models on how to capture user intentions. These technologies and ideas can more effectively filter unintentional information in the user behavior sequence, modeling user's high-order intention [13]. Therefore, introducing the idea of sequential recommendation into CTR prediction is undoubtedly an exciting idea, which can bring new ideas on how to use the user historical behavior sequence for CTR prediction.

The quality of the introduced sequential recommendation model determines the final prediction effect, but in fact, there are still some problems with sequential recommendation: First, aimless browsing by users is a common phenomenon in many recommendation scenarios (e-commerce, music, video stream), the user may be in status with no intention at present. In this status, the user's browsing is a process of exploration, and the user will tend to browse things that he/her has not seen and is interested in. In other words, users do not deliberately follow or look for something, they just browse casually. Figure 1 is a simple example to illustrate the difference between a user in an intentional status and a status without an intention. For $user_1$, his interaction history in the previous period is

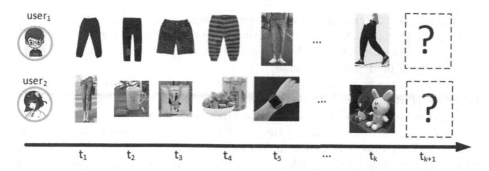

Fig. 1. User's history behavior in different status.

items of pants type. It can be seen that he has a strong intention to browse or even buy pants, so the recommendation system should recommend him another pair of pants at the next moment. For $user_2$, she has a rich history of interactions in the past. In addition to browsing pants, she also clicked and browsed many different types of items. There seems to be no obvious connection between these items. Therefore, if the recommendation system recommends a pair of pants or an apple watch to her according to the intentional state at the next moment, it would not be so wise. This example is extreme, because it only reflects the situation in a single status of the user. In real life, users are likely to be a mixture of two status when browsing items, so it is necessary to comprehensively consider how to more accurately model the user's interests in the two status. Second, when extracting user intent, it is often only to model the relationship between the items that have interacted, without considering the connection between the candidate set item and the behavior sequence, and only using a simple similarity score function to represent the user's preference for the candidate set item [6,8,13]. In fact, it's not only the user's behavior that has a strong order and inertia, there is a certain degree of sequential connection between items, because after browsing an item, the user will also be interested in the related items of this item. If you do not consider the degree of correlation between the candidate set items and the behavior sequence, you will undoubtedly lose a lot of useful information.

First of all, in order to solve the first problem, this paper introduces the concept of user browsing status, which is used to characterize whether the user has an intention in the current behavior sequence. The user status is divided into Discover and Intent. Discover indicates that user is in a status with no obvious intention, and Intent means that user has obvious intention. Then, this paper proposes a framework called User Status Recognition Framework (USRF), which contains four parts, namely User Status Classification Module, Discover Prediction Module, Intent Prediction Module, Final Prediction Module. User Status Recognition Module is used to capturing the user's weight for the two status, the two Prediction Modules use different models to model user interests in different status. Finally, the results of the first three modules are combined in the

Final Prediction Module to generate the final recommendation result. It is worth mentioning that the USRF can deal with Sequential Recommendation and CTR prediction. The similarities and differences between sequential recommendation and CTR will be detailed in Sect. 3.

Then, this paper implements a model applied to the CTR predition problem based on USRF, and uses the attention mechanism in the model to model the association between the items in the candidate set and the items in the behavior sequence. This method solves the second problem mentioned earlier in this paper. In User Status Recognition Module, we uses the attention mechanism to model the complex relationship between users and behavior sequences, and then captures the user's state weight. In both prediction modules, the relationship between the candidate item and the interacted item is modeled. Finally, the final score of the candidate item is obtained by combining the user's status weight in the Final Prediction Module.

The main contributions of this paper are as follows:

1. In order to solve the problem that the user behavior is regarded as an independent signal in the CTR estimation, sequential recommendation's idea is introduced to accurately model the user's historical behavior to perform CTR estimation.
2. In order to solve the situation where the user has no obvious intention, the concept of user browsing status is introduced. The user browsing status is divided into Discover and Intent, and a framework for capturing user status weights and modeling user interests separately is proposed. The framework is called User Status Recognition Framework (USRF).
3. In the model based on the realization of USRF, the attention mechanism was used to model the relationship between the candidate set item and the interacted item, and the model was used for CTR prediction. Experiments were performed on three different scene data sets. Both AUC and F1-score indicators have achieved good results, proving the advantages of USRF.

2   Related Work

The research content of this paper mainly involves two directions, CTR estimation and sequential recommendation.

2.1   Click-Though Rate Prediction

Driven by the growing demand of the Internet, the CTR estimation model has developed very rapidly. From the common logistic regression (LR) [14] before 2010, it has evolved to a factorization machine (FM) [1], Gradient Boosting Decision Tree (GBDT) [15]. With the popularity of deep learning after 2015, various deep learning-based model architectures have emerged endlessly. The first is that FM [1] introduces hidden vectors, so that feature crossing becomes efficient, and then introduces the concept of Field-aware FM (FFM) [3], making the model more expressive. The Deep Crossing [16] proposed by Microsoft in 2016 is the most typical and basic model of the deep learning CTR model, and the FNN [17] proposed later uses the FM hidden vector as the embedding of users and items. PNN [18] adds a Product layer between the embedding layer and the fully connected layer, which can capture different cross-information and enhance the model's ability to represent different data modes. The Wide&Deep [19] model proposed by Google introduced MLP to enhance the memory ability of the model. DeepFM [2] and Deep&Cross [20] are both the result of Wide&Deep improvement. The original Wide part was replaced with a new network structure. And the NFM [21] proposed by He et al. improved the Deep part, and replaced the feature crossing part of FM with Bi-interation Pooling. The AFM [4] proposed by He et al. started to introduce the attention mechanism into FM, DIN [10] also added the attention mechanism to its own deep learning CTR model, and DIEN [12] began to process the user's sequence behavior based on DIN. Most of the existing researches on CTR models regard the user's historical behavior as an independent signal, and there are relatively few studies focusing on the order relationship of user behavior, and the research is not deep.

2.2 Sequential Recommendation

Sequential recommendation is a sub-task of the recommendation system. The traditional content-based and collaborative filtering recommendation algorithm have defects in the sequence data: each item is independent of each other, and continuous preference information of item in the session cannot be modeled. With the development of deep learning, especially RNN in natural language processing (NLP) area, people are increasingly aware of the advantages of models such as RNN in processing serialized data. Starting from the RUE4Rec [7] in 2016, sequence recommendation has entered a stage of rapid development, and many researchers have proposed various sequence recommendation models in recent years. Veronika Bogina et al. [22] proposed in the model proposed in 2017 to take into account the length of time the user stayed in the item in the session. Quadrana, Massimo et al. [23] proposed a hierarchical RNN model. Compared with the previous work, the user's personal interest changes in the session can be described, and the user's personalized session recommendation can be made. Tan et al. [24] introduced four optimization methods in GRU4REC, including data augmentation, model pre-training, use of privileged information and output embedding. In the past two years, more and more scholars have combined sequence recommendations with other work. For example, the attention mechanism [6, 8, 23], graph neural network(GNN) [25], BERT [26], convolutional neural network (CNN) [27] and so on.

3   Proposed Method

This chapter first introduces the similarities and differences between the sequential recommendation task and the CTR prediction task, and how to introduce

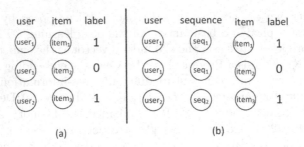

Fig. 2. CTR's sample VS sequential recommendation's sample.

the sequential recommendation into the CTR prediction, and then introduces USRF and the model based on the USRF proposed in this paper.

3.1   CTR vs Sequential Recommendation

In simple terms, sequence recommendation is to capture the user's dynamic intention from user's recent historical behavior sequence, and then use user's intention to calculate candidate items' score, select the K highest-scoring items from the candidate set for Top-k recommendation, also called next -item recommended. For the CTR prediction problem, the purpose is to model the complex relationship between a user and a target item, calculate the degree of user preference for the target item, that is, the click probability.

In other words, for sequence recommendation, if the number of items in the candidate set is reduced to 1, sequence recommendation is also calculating the degree of user preference for a target item. Therefore, it is feasible to introduce the idea of sequence recommendation into the problem of CTR prediction, and the rich processing methods and creative ideas on sequence of user behavior in the field of sequence recommendation can also bring new ideas to the problem of CTR prediction.

Of course, there is a certain difference in data processing between CTR prediction problem and sequence recommendation. The classic algorithm NCF [9] in CTR field is used as an example to briefly introduce the data processing methods. As shown in Fig. 2, Fig. 2(a) is an example of NCF sample data. A sample can be represented as a triple $(user, item, label)$, where $user \in U = \{u_1, u_2, u_3, , u_{|U|}\}$, represents user index, $item \in I = \{i_1, i_2, i_3, , i_{|I|}\}$, represents item index, $label \in \{0, 1\}$, $item$ is the target item to be predicted, $label$ indicates whether user clicked the item, 0 indicates no click, and 1 indicates click. Figure 2(b) is an example of sample data for the implementation model of this paper. A sample can be represented as a quaternion $(user, sequence, item, label)$, where $sequence = \{item_1, item_2, , item_t\}$, which is the item that the user interacted with at the previous t moments, item is the item to be predicted at $t + 1$. The label indicates whether user clicked the item.

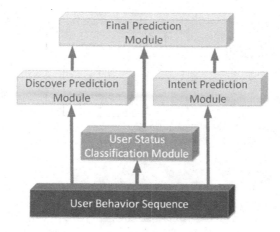

Fig. 3. User Status Recognition Framework.

3.2 Framework and Model

Framework. The User Status Recognition Framework (USRF) proposed in this paper is shown in Fig. 3. This framework contains four modules, namely User Status Classification Module, Discover Prediction Module, Intent Prediction Module, and Final Prediction Module. User Status Recognition Module is used to capture weights of the two user status, the two prediction modules respectively use different models to model user interests in different status. Finally, the results of the first three modules are combined in the Final Prediction Module to generate final recommendation result. In Sect. 3.1, we have demonstrated the similarities and differences between CTR prediction and sequence recommendation, mainly due to the difference between the number of items in candidate set. Therefore, by adjusting the processing method of the candidate set, this framework can be used for CTR and sequence recommendation.

Model. This paper implements a model based on USRF for CTR prediction, as shown in Fig. 4. Give a $user_i \in U$, user's historic behavior is denoted as $sequence_i = \{item_1, item_2, ..., item_t\}$, and the target item is denoted as $item_{target} \in I$. $(user_i, sequence_i, item_{target})$ is passed to Embedding Layer as the input of model, and user embedding vector denotes as $u_i \in \mathbb{R}^d$, user sequence matrix denotes as $seq_i \in \mathbb{R}^{t \times d}$, target item embedding vector denotes as $i_{target} \in \mathbb{R}^d$. This paper uses the attention mechanism to capture the user's preference for the behavior sequence items. First, u_i and seq_i are passed into Attention Net, and the attention weight a_k of $user_i$ to $item_k \in sequencen_i$ is calculated as follows :

$$a_k = W_0(\sigma(W_u \cdot u_i + W_i \cdot i_k + b_0)) \tag{1}$$

Where $W_0, W_u, W_i \in \mathbb{R}^{d \times d}, b_0 \in \mathbb{R}^d$ are learnable parameters, $\cdot$ is the inner product, σ is the sigmoid function.

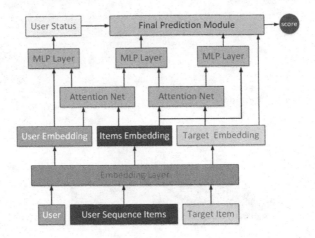

Fig. 4. Detail of MyModel.

After calculating all the attention weights, perform a weighted average on $sequence_i$ to get the user's general intention presentation vector $v_{gi} \in \mathbb{R}^d$, the formula as follows:

$$v_{gi} = \frac{1}{t} \sum_{k=1}^{t} a_k \cdot i_k \tag{2}$$

Then pass v_{gi} to a fully connected layer to capture the weight of the user's two status:

$$s = f(W_s \cdot v_{gi}) \tag{3}$$

Where $W_s \in \mathbb{R}^{2 \times d}$ are learnable parameters, $f(\cdot)$ is a non-linear function, and $s = [status_1, status_2]$.

The above part implements the User Status Classification Module of USRF. When implementing two prediction modules, this paper uses attention mechanism to model the association between target item and history items. First, i_{target} and seq_i are passed into Attention Net, and the attention weight a_p of $item_{target}$ to $item_p \in sequence_i$ is calculated as follows:

$$a_p = W_1(\sigma(W_t \cdot i_{target} + W_j \cdot i_p + b_1)) \tag{4}$$

Where $W_1, W_t, W_j \in \mathbb{R}^{d \times d}, b_1 \in \mathbb{R}^d$ are learnable parameters After calculating all the attention weights, perform a weighted average on $sequence_i$ to get the correlation vector $v_{ir} \in \mathbb{R}^d$, the formula as follows:

$$v_{ir} = \frac{1}{t} \sum_{p=1}^{t} a_p \cdot i_p \tag{5}$$

For users in the Discover status, there is no clear intention. User's behavior at the next moment will be mainly affected by the user's behavior inertia, that is, the behavior at the previous moment will have the greatest impact on the next moment. So this paper use last item in user behavior sequence to model user interest, stitching $item_t$ and v_{ir} into a fully connected layer to get the user interest vector $v_{di} \in \mathbb{R}^d$ in Discover status, the formula is as follows:

$$v_{di} = f(W_{di} \cdot concat(i_t, v_{ir})) \tag{6}$$

Where $W_{di} \in \mathbb{R}^{d \times 2d}$ are learnable parameters.

However, users in the Intent status have clear attention, so this paper combines user general intention presentation vector v_{gi} and v_{ir} into a fully connected layer to get user interest vector $v_{ii} \in \mathbb{R}^d$ in the Intent status. The formula is as follows:

$$v_{ii} = f(W_{ii} \cdot concat(v_{ir}, v_{gi})) \tag{7}$$

Where $W_{di} \in \mathbb{R}^{d \times 2d}$ are learnable parameters.

The above part implement USRF's Discover Prediction Module and Intent Prediction Module. In the model's Final Prediction Module, final score calculation of target item is divided into two parts. First, v_{di} and v_{ii} are used to calculate the user's score in different status with target item. The calculation formulas of the score of the target item $item_{target}$ are:

$$score_d = v_{di}^T \cdot i_{target} \tag{8}$$

$$score_i = v_{ii}^T \cdot i_{target} \tag{9}$$

Then the score vector $score = [score_d, score_i]$ is multiplied with the user's state weight vector s to get final score of target item:

$$score_f = \sigma(s \cdot score^T) \tag{10}$$

We optimize the proposed model by using a standard mini-batch gradient descent on the cross-entropy loss:

$$L = -\sum_{i=1}^{m} y_i log(\hat{y}_i) + (1 - y_i)log(1 - \hat{y}_i) \tag{11}$$

where y_i denote the label of sample i, $\hat{y}_i$ denote the prediction probability.

4   Experiments

In this section, we evaluate the proposed model with the state-of-the-art methods on three real-world datasets.

4.1   Datasets

The proposed model is evaluated on three real-world datasets from various domains: Kaggle's retailRocket datasets, movielens-1m datasets and Amazon's digitalMusic datasets. To ensure the experimental results, there are some preprocessing for the dataset. First, we filter out the items that appear less than 5 and

Table 1. Statistics of the dataset.

	retailRocket	ml-1m	digitalMusic
User	21716	6025	5260
Item	52074	3282	22140
Train sample	689766	1065414	79556
Test sample	43432	12010	10520

the users whose interaction records are less than 10 (for movielens-1m datasets, the number is 3 and 5, for digitalMusic datasets, the number is 3 and 7). In terms of data enhancement, session length is fixed as $t = \{5, 7, 10\}$, and then the $t + 1$ interaction as a label. A user's action sequence can generate several samples, which is more conducive to training model. Same as NCF [9], use leave-one-out to split dataset. For each user, leave the last sample for test, and other samples for train. Since the samples in the data set are all positive samples, sampling is required. For each positive sample we randomly sample one negative sample. The statistics of the three data sets after preprocessing are as shown in Table 1.

4.2   Compared Methods

We compare our model with the following competitive models:

GRU4Rec [7]: An RNN based deep learning model for session-based recommendation, which consists of GRU units, it utilizes session-parallel mini-batch training process and also employs ranking-based loss functions during the training.

NARM [6]: An RNN based state-of-the-art model which employs attention mechanism to capture main purpose from the hidden states and combines it with the sequential behaviors final representation to generate recommendations.

STAMP [8]: This is a model proposed in 2018. This model uses MLP and attention mechanism to extract the long-term and short-term intent of the user in the session for the next-item recommendation.

NueMF [9]: NueMF is a model based on a generic framework NCF, and it can express and generalize matrix decomposition. In order to improve the ability of nonlinear modeling, the model uses multi-layer perceptron to learn the user-item interaction functions.

This paper we take the task as Click-Through Rate Prediction, thus we use AUC and F1-score as the evaluation metrics. AUC measures the probability that a positive instance will be ranked higher than a randomly chosen negative one. F1-score is the harmonic mean of precision and recall. These metrics summarize a model's performance from different aspects [28]. In this paper, five experiments were conducted for each dataset, and then the average value was taken as the final result.

Table 2. Comparison to baselines.

Model/dataset	retailRocket		ml-1m		digitalMusic	
	AUC	F1-socre	AUC	F1-socre	AUC	F1-socre
GRU4Rec	0.8119	0.7453	0.9068	0.8287	0.6856	0.6420
NARM	0.8473	0.7631	0.9183	0.8376	0.6208	0.5869
STAMP	0.8422	0.7627	0.9084	0.8206	**0.7102**	0.6357
NueMF	0.8593	0.7724	0.8814	0.7924	0.6757	0.6233
Mymodel	**0.8888**	**0.8001**	**0.9221**	**0.8445**	0.7086	**0.6709**

4.3 Results

This section compares and analyzes the results of our model and baseline method, and then analyzes the effectiveness of the proposed user state classification and modeling of the relationship between candidate items and behavior sequences.

Comparison to Baselines. As shown in Table 2, the performance of the model implemented base on USRF is better than the baseline method on three different domain datasets. First, the reasons why the model proposed in this paper is superior to NueMF may be: (1) the idea of sequence recommendation is introduced, the user's behavior history is complicatedly modeled, the user's interest is dynamically characterized, and (2) compared with MLP is used to model the relationship between users and items. The model in this paper uses the attention mechanism to model the relationship between candidate items and user behaviors to better express the potential interests of users. Second, the common reasons why the model proposed in this paper is superior to the three sequential recommendation models may be: (1) the idea of user status classification is introduced to solve the error that may be caused by modeling user interest in unintended status, (2) modeling the association between candidate items and historical behavior items.

Effectiveness of User Status Classification. This paper divides the model implemented based on USRF into three models to prove the effectiveness of classifying user status. The results are shown in Table 3. Mymodel-discover and Mymodel-intent are models implemented using only a single prediction module respectively. The experiment proves that the effect of the model is improved after classifying the user status and considering the two status comprehensively on three datasets.

Effectiveness of Modeling the Association. This paper also implement a model that not model the association between candidate items and historical behavior items, to prove the effect of modeling the association, we call this model as Mymodel-without. The results are shown in Table 3, the experiment proves that the effect of the model is improved after modeling the association between candidate items and historical behavior items on three datasets.

Table 3. Effectiveness of user status classification and modeling the association.

Model/Dataset	retailRocket		ml-1m		digitalMusic	
	AUC	F1-socre	AUC	F1-socre	AUC	F1-socre
Mymodel-discvoer	0.8718	0.7904	0.9142	0.8375	0.7009	0.6671
Mymodel-intent	0.8767	0.7966	0.9154	0.8360	0.7011	0.6654
Mymodel-without	0.8786	0.7879	0.9121	0.8312	0.7039	0.6683
Mymodel	**0.8888**	**0.8001**	**0.9221**	**0.8445**	**0.7086**	**0.6709**

5 Conclusion

In this paper, we introduce sequential recommendation into the field of CTR prediction, and propose the concept of user browsing status. The user browsing status is divided into unintentional status (Discover) and intentional status (Intent). The User Status Recognition Framework (USRF) is proposed to use different methods mining user interest in different user status. Then this paper implements a simple model based on USRF. In the model, the association between candidate items and historical behavior items is modeled to make up for the lack of previous sequential recommendation work. The performance of the model and effectiveness of user status classification and modeling the association are proved on three different scene datasets. In future work, we will follow this idea to conduct a more in-depth analysis of the user's unintentional status and intentional status, and implement models that are suitable for the two status. At the same time, further research on the method of user status classification can more accurately capture the user's current status weight.

References

1. Rendle, S.: Factorization machines. In: 2010 IEEE International Conference on Data Mining, pp. 995–1000. IEEE (2010)
2. Guo, H., Tang, R., Ye, Y., et al.: DeepFM: a factorization-machine based neural network for CTR prediction. arXiv preprint arXiv:1703.04247 (2017)
3. Juan, Y., Zhuang, Y., Chin, W.S., et al.: Field-aware factorization machines for CTR prediction. In: Proceedings of the 10th ACM Conference on Recommender Systems, pp. 43–50. ACM (2016)
4. Xiao, J., Ye, H., He, X., et al.: Attentional factorization machines: learning the weight of feature interactions via attention networks. arXiv preprint arXiv:1708.04617 (2017)
5. Wang, S., Cao, L., Wang, Y.: A survey on session-based recommender systems. arXiv preprint arXiv:1902.04864 (2019)
6. Li, J., Ren, P., Chen, Z., et al.: Neural attentive session-based recommendation. In: Proceedings of the 2017 ACM on Conference on Information and Knowledge Management, pp. 1419–1428 ACM (2017)
7. Hidasi, B., Karatzoglou, A., Baltrunas, L., et al.: Session-based recommendations with recurrent neural networks. arXiv preprint arXiv:1511.06939 (2015)

8. Liu, Q., Zeng, Y., Mokhosi, R., et al.: STAMP: short-term attention/memory priority model for session-based recommendation. In: Proceedings of the 24th ACM SIGKDD International Conference on Knowledge Discovery Data Mining, pp. 1831–1839. ACM (2018)

9. He, X., Liao, L., Zhang, H., et al.: Neural collaborative filtering. In: Proceedings of the 26th International Conference on World Wide Web, pp. 173–182, International World Wide Web Conferences Steering Committee (2017)

10. Zhou, G., Zhu, X., Song, C., et al.: Deep interest network for click-through rate prediction. In: Proceedings of the 24th ACM SIGKDD International Conference on Knowledge and Discovery Data Mining, pp. 1059–1068. ACM (2018)

11. Feng, Y., Lv, F., Shen, W., et al.: Deep session interest network for click-through rate prediction. arXiv preprint arXiv:1905.06482 (2019)

12. Zhou, G., Mou, N., Fan, Y., et al.: Deep interest evolution network for click-through rate prediction. In: Proceedings of the AAAI Conference on Artificial Intelligence, pp. 33, pp. 5941–5948 (2019)

13. Ma, C., Kang, P., Liu, X.: Hierarchical gating networks for sequential recommendation, arXiv preprint arXiv:1906.09217 (2019)

14. Wright, R.E.: Logistic regression (1995)

15. Roe, B.P., Yang, H.J., Zhu, J., et al.: Boosted decision trees as an alternative to artificial neural networks for particle identification. Nucl. Instrum. Methods Phys. Res., Sect. A 543(2–3), 577–584 (2005)

16. Shan, Y., Hoens, T.R., Jiao, J., et al.: Deep crossing: web-scale modeling without manually crafted combinatorial features. In: Proceedings of the 22nd ACM SIGKDD International Conference on Knowledge Discovery and Data Mining, pp. 255–262, ACM (2016)

17. Zhang, W., Du, T., Wang, J.: Deep learning over multi-field categorical data. In: Ferro, N., et al. (eds.) ECIR 2016. LNCS, vol. 9626, pp. 45–57. Springer, Cham (2016). https://doi.org/10.1007/978-3-319-30671-1_4

18. Qu, Y., Cai, H., Ren, K., et al.: Product-based neural networks for user response prediction. In: 2016 IEEE 16th International Conference on Data Mining (ICDM), pp. 1149–1154. IEEE (2016)

19. Cheng, H.T., Koc, L., Harmsen, J., et al.: Wide and deep learning for recommender systems. In: Proceedings of the 1st Workshop on Deep Learning for Recommender Systems, pp. 7–10. ACM (2016)

20. Wang, R., Fu, B., Fu, G., et al.: Deep and cross network for ad click predictions. In: Proceedings of the ADKDD 2017, p. 12. ACM (2017)

21. He, X., Chua, T.S.: Neural factorization machines for sparse predictive analytics. In: Proceedings of the 40th International ACM SIGIR Conference on Research and Development in Information Retrieval, pp. 355–364. ACM (2017)

22. Bogina, V., Kuflik, T.: Incorporating dwell time in session-based recommendations with recurrent neural networks, RecTemp@ RecSys 2017)

23. Quadrana, M., et al.: Personalizing session-based recommendations with hierarchical recurrent neural networks. In: Proceedings of the Eleventh ACM Conference on Recommender Systems. ACM (2017)

24. Tan, Y.K., Xu, X., Liu, Y.: Improved recurrent neural networks for session-based recommendations. In: Proceedings of the 1st Workshop on Deep Learning for Recommender Systems. ACM (2016)

25. Wu, S., et al.: Session-based recommendation with graph neural networks. In: Proceedings of the AAAI Conference on Artificial Intelligence, vol. 33 (2019)

26. Sun, F., et al.: BERT4Rec: sequential recommendation with bidirectional encoder representations from transformer, arXiv preprint arXiv:1904.06690 (2019)

27. Tang, J., Wang, K.: Personalized top-N sequential recommendation via convolutional sequence embedding. In: WSDM, pp. 565–573. ACM (2018)
28. Yu, Z., Lian, J., Mahmoody, A., et al.: Adaptive user modeling with long and short-term preferences for personalized recommendation. In: Proceedings of the 28th International Joint Conference on Artificial Intelligence, pp. 4213–4219. AAAI Press (2019)

Author Index

Printed in the United States
By Bookmasters